First published 2012 by Pisces Publications for St Helena Nature Conservation Group.
Pisces Publications is the imprint of NatureBureau.

British Library-in-Publication Data
A catalogue record for this book is available from the British Library.

ISBN 978 1 874357 51 3

Designed by NatureBureau, 36 Kingfisher Court, Hambridge Road, Newbury, Berkshire RG14 5SJ
www.naturebureau.co.uk

Companion books published 2012 by Pisces Publications

Flowering Plants & Ferns of St Helena ISBN 978 1 874357 52 0
Lichens of St Helena ISBN 978 1 874357 53 7

Cover photographs
FRONT *Pleurozia gigantea*, growing with *Macromitrium microstomum* on a black cabbage tree, Diana's Peak National Park (Phil Lambdon); *Cylindrocolea helenae, Macromitrium urceolatum, Philonotis heleniana* (Martin Wigginton)
BACK *Campylopus arcuatus* and *Kurzia nemoides* (left) and *Symphyogyna brasiliensis* (right) on an earthy bank (Andrew Darlow)

Contents

Acknowledgements

The author extends grateful thanks to all those who have contributed to this guide by generously giving of their time and expertise, and without whom this book would not have been possible. On St Helena, I am indebted to Andrew Darlow and Rebecca Cairns-Wicks for instigating the project, and to the St Helena Nature Conservation Group for supporting it; Rebecca Cairns-Wicks also provided logistical help during field surveys and commented on the manuscript. Further thanks to staff at the Conservation Unit, Scotland; Annalea Beard for obtaining *Physcomitrium* capsules; and to the St Helena Government for permission to collect specimens. Cartographic data was obtained from the island's SHEIS GIS system, courtesy of Len Coleman, the St Helena Government GIS manager.

Funding for field surveys and the production of this guide is gratefully acknowledged from the UK Government's Overseas Territories Environment Programme (a joint programme of the Foreign and Commonwealth Office and the Department for International Development). Continuing support for the project and practical assistance with shipping, by Mrs Kedell Worboys, St Helena Government's UK representative, is greatly appreciated.

The following institutions and the curators of their herbaria are thanked for the loan of specimens or other information: British Museum (Natural History), London; Royal Botanic Garden, Edinburgh; National Museum of Wales, Cardiff; Göttingen University; Conservatoire et Jardin Botanique de la Ville de Genève; Netherlands National Herbarium, Leiden; New York Botanical Gardens; Farlow Herbarium, Harvard University. My particular thanks to L.T. Ellis, Curator of Bryophytes at the British Museum (Natural History), for providing facilities on many occasions for the author to study the historical St Helena collections, and for facilitating loans from other institutions.

The author is indebted to the following people for help in species identification: Bill Buck (*Lepidopilidium*), Jochen Heinrichs (*Marchesinia*), David Holyoak (*Bryum*), Mikhail Ignatov (*Sainthelenia*), Howard Matcham (*Pottiaceae*), Maria Gallego Morales (*Pseudocrossidium*), Frank Müller (*Aureolejeunea rotalis*), Tamás Pócs (*Lepidozia*), Jiri Váňa (*Cylindrocolea*), and Ida Bruggemann-Nannenga for much help identifying *Fissidens* species.

As well as contributing some additional data on the bryophytes of St Helena, my grateful thanks are due to Phil Lambdon and Andrew Darlow for steering this project through the editorial stages, and to Andrew for producing the distribution maps. Lastly, I thank Peter Creed and staff at NatureBureau for their assiduous work compiling the colour plates and designing the book.

Foreword

The small and isolated island of St Helena lies in the southern Atlantic Ocean, about mid-way between Angola to the east and Brazil to the west, with the nearest land, Ascension Island, some 1,300 km to the north-west. The island rises steeply from the sea on all sides, and several cliffs are more than 400 m high. The highest ground forms a steep-sided ridge 700–820 m high, and much of the island is dissected by deep valleys, with very little level ground except locally in the east.

The island is home to many rare and endemic species of plants and animals, the importance of which have long been recognised. The mosses and liverworts remained little known, however, with information largely confined to scientific journals, and difficult to access. The main aim of publishing a guide was, therefore, to bring these plants more within the easy reach of students and naturalists. It is hoped that this book will stimulate an interest in this group of plants, and encourage their study by islanders and visitors alike. A great deal remains to be discovered about the moss and liverwort flora of St Helena, and species new to the island are certain to be found.

The island of St Helena

The main places of bryological interest mentioned in the text are indicated, together with major settlements (in capital letters)

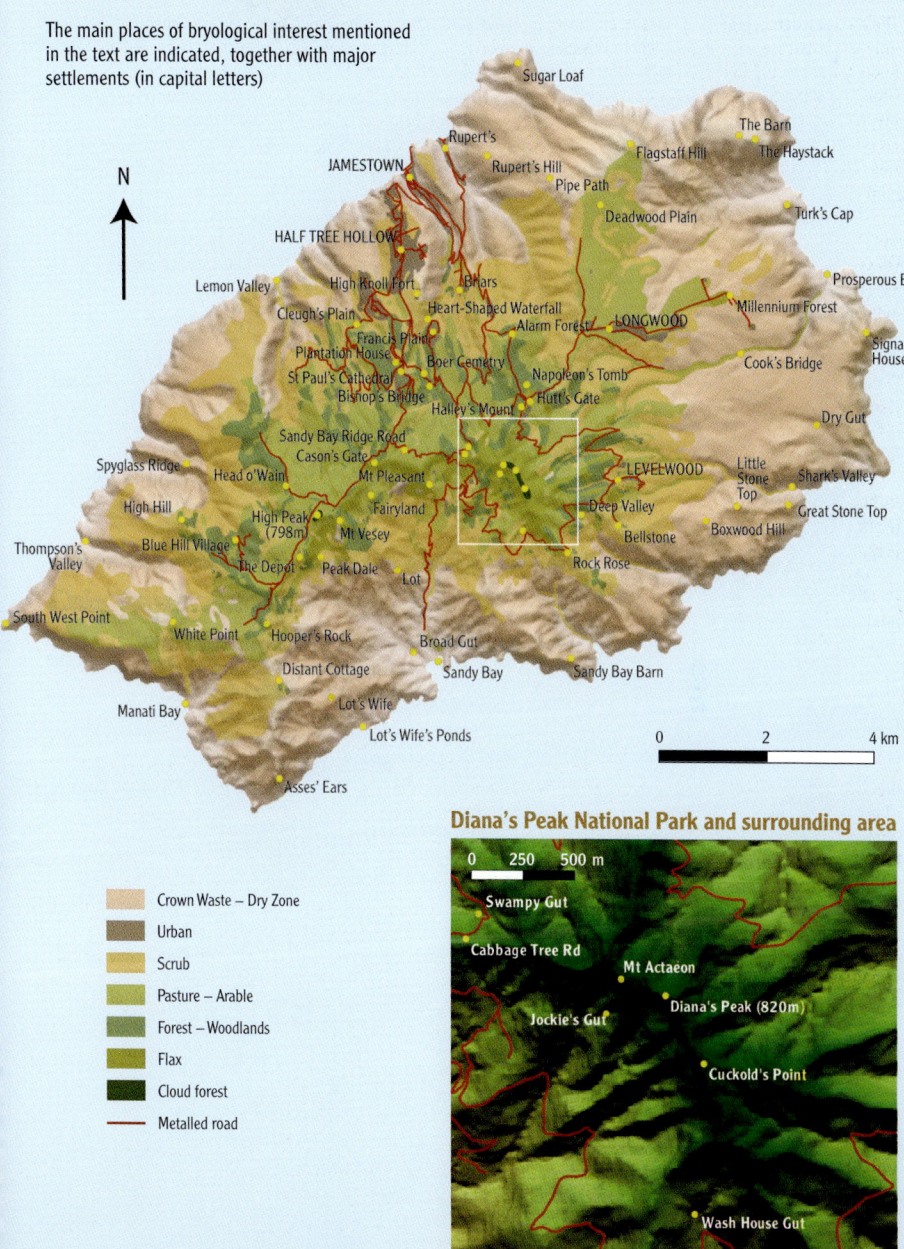

N

Sugar Loaf

Rupert's
JAMESTOWN
Rupert's Hill
Pipe Path
Flagstaff Hill
The Barn
The Haystack

HALF TREE HOLLOW
Deadwood Plain
Turk's Cap

Lemon Valley
High Knoll Fort
Briars
Heart-Shaped Waterfall
Alarm Forest
LONGWOOD
Millennium Forest
Prosperous Bay

Cleugh's Plain
Francis Plain
Plantation House
St Paul's Cathedral
Boer Cemetery
Napoleon's Tomb
Bishop's Bridge
Halley's Mount
Hutt's Gate
Cook's Bridge
Signal House

Spyglass Ridge
Head o'Wain
Sandy Bay Ridge Road
Cason's Gate
Mt Pleasant
LEVELWOOD
Little Stone Top
Shark's Valley
Dry Gut

High Hill
High Peak
(798m)
Fairyland
Mt Vesey
Deep Valley
Great Stone Top
Boxwood Hill

Thompson's Valley
Blue Hill Village
The Depot
Peak Dale
Lot
Bellstone
Rock Rose

South West Point
White Point
Hooper's Rock
Broad Gut
Sandy Bay
Sandy Bay Barn

Manati Bay
Distant Cottage
Lot's Wife
Lot's Wife's Ponds

Asses' Ears

0   2   4 km

Crown Waste – Dry Zone
Urban
Scrub
Pasture – Arable
Forest – Woodlands
Flax
Cloud forest
Metalled road

Diana's Peak National Park and surrounding area

0   250   500 m

Swampy Gut
Cabbage Tree Rd
Mt Actaeon
Jockie's Gut
Diana's Peak (820m)
Cuckold's Point
Wash House Gut

Introduction

The bryophytes, referred to in common parlance as 'mosses' actually consist of several different groups of green plants which occupy an intermediate evolutionary position between the algae and the 'higher' plants (which include flowering plants and ferns); they include the true mosses as well as the liverworts and hornworts. They are ancient groups, with a fossil record going back to the Devonian period, appearing at least two hundred million years before the flowering plants. The total number of bryophyte species worldwide is relatively small, probably fewer than 20,000.

Most bryophytes are small, ranging when fully grown from less than 1 mm to a few cm long/high. Some can attain greater sizes, however, and tropical *Dawsonia* species can grow unsupported to 80 cm high (although they are usually shorter). Pendent species can produce very long, hanging shoots, and a few aquatic species can reach considerable lengths floating in water. Bryophytes are found in most habitats, but none occur in marine environments. Some species are generalists that can be found in a wide range of situations on a variety of substrates, whilst others are much more exacting, being confined to a particular habitat, even a specific type of bark, soil or rock. They are often thought of as plants of moist habitats, but many species are found in, or confined to, very dry places. Some can survive long periods of desiccation in desert conditions, shrivelling and drying almost completely, yet reviving quickly with the onset of rain and resuming photosynthesis.

The St Helena bryophyte flora

The bryophyte flora of St Helena is small, with only about 110 species currently known, though others are certain to be found. The remoteness of the island and its relatively recent volcanic origin from the Mid-Atlantic Ridge only 12–14 million years ago, when the Atlantic Ocean was almost as wide as it is today, means that all, or almost all, natural colonisations must have had their origin in spores carried by the wind over the ocean, before germinating in suitable habitats. Many of the arrivals are recognised as the same species that occur in Africa, the Americas and elsewhere. Others have evolved into new forms found nowhere else (i.e., endemic), and St Helena is known to be an important centre of bryophyte endemism. As well as indigenous species, there are also several that have almost certainly, or probably, been introduced into the island by man. Some species, notably *Pseudoscleropodium purum*, are known to have been used as packing material for imported plants. Others might have arrived accidentally with commercial or decorative plants introduced over the past few centuries, or perhaps have originated from spores or fragments in non-sterile peat imported for use in nurseries and gardens.

The first bryophytes recorded from the island were *Pleurozia gigantea* and *Macromitrium urceolatum*, collected by Archibald Menzies in 1795, the latter species published with a beautiful illustration, 25 years later by J.D. Hooker in his *Musci Exotici*. William Burchell, who was resident on the island between 1805 and 1810, made the first significant collections of bryophytes, and although a succession of other European botanists visited the island during the 19th century, rather few bryophyte collections resulted from most of those visits. However, important collections were made by J.D. Hooker who visited and collected on the island in 1839 and 1844, and J.C. Melliss collected some bryophytes in the latter half of the century. W. Mitten, who did not visit the island, identified bryophytes collected by Melliss and other Victorian botanists, and his important findings were published in 1875 in Melliss's book that describes the island, including its natural history. There are very few records from the 20th century, but a few collections were made in 1992 by M. Karisch (identified by F. Müller, 1999) and by A. Eastwood in 1999, by which time about 50 bryophytes were known from the island. A major survey of St Helena bryophytes was carried out by the author in October and November 2005, during which time extensive small collections were made, and many additions made to the flora. The present guide is based largely on the information gathered during the 2005 survey, together with some subsequent records and observations made by R. Cairns-Wicks and P. Lambdon.

Much remains to be learned about the bryophyte flora of the island, including the habitats of the species and their geographical distribution. Some bryophytes can only be identified to the genus level at the present time (e.g. *Riccardia*, *Cephaloziella*, *Isopterygium*), and the identities of some other species need to be investigated further. The rich and important bryoflora of the native cloud forest communities is reasonably well known, but much less is known about the bryophyte communities and the distributions of the species at middle and low altitudes on the island.

Bryophyte habitats and distribution

The bryologically richest parts of the island are found at altitudes above 600 m, especially in the cloud forest habitat in Diana's Peak National Park and at High Peak. Other notable communities can be found elsewhere along the Central Ridge, including at the Depot and Hooper's Rock. Some parts of the Ridge have not been explored, and it seems likely that additional areas of interest could be found elsewhere, perhaps around Cason's Gate and Mount Vesey. About 60 species of bryophyte are found in Diana's Peak National Park and along the Central Ridge, of which half appear to be confined to this area. Of the presumed endemics, 22 of the 26 species occur here, of which ten are found only in the native-rich cloud forest communities and adjacent natural habitats. These are: *Aureolejeunea rotalis, Cololejeunea dianae, Dendroceros adglutinatus, Kurzia nemoides, Lejeunea sanctae-helenae, Tylimanthus anisodontus, Fissidens chioneurus, Lepidopilidium crispifolium, Sematophyllum erythrocaulon* and *Sphagnum helenicum.*

Trunks and branches of native trees and shrubs (especially the cloud forest species: tree fern, black cabbage tree and whitewood), support a varied epiphytic flora, including *Chiloscyphus humistratus, Radula fulvifolia, Pleurozia gigantea* (locally), *Macromitrium microstomum, Sematophyllum erythrocaulon* and many species of the Lejeuneaceae (*Aureolejeunea rotalis, Cololejeunea* spp., *Colura* spp. and *Microlejeunea africana*). Bryophyte communities are often well-developed on shaded, peaty or rocky banks, and typically include *Adelanthus decipiens, Bazzania praerupta, Campylopus arcuatus, Fissidens chioneurus, Kurzia nemoides, Mnioloma fuscum, Sematophyllum erythrocaulon, Symphyogyna brasiliensis* and *Syrrhopodon gaudichaudii.*

At high altitudes, earthy banks by roads and paths, often shaded by overhanging grass and herbs, have a varied bryoflora. Species in these habitats include *Adelanthus decipiens, Chiloscyphus humistratus, Dicranella proscripta, Entosthodon* sp., *Lepidopilidium pallidifolium, Radula fulvifolia* and *Riccardia* spp.

Much less is known about bryophyte habitats and the distribution of species outside the national park and the Central Ridge, and many areas at middle altitudes have not been explored bryologically. However, bryophytes are to be found in most habitats, though are nearly absent from some, such as *Eucalyptus* woodland and coarse-grass pasture. Mid-altitude roadside banks often have a varied bryoflora: frequent species include *Bryum dichotomum, Dicranella proscripta, Phaeoceros carolinianus, Philonotis helenicus, Pseudoscleropodium purum* and *Cephaloziella* spp. Large Caffra thorn and other trees often support a characteristic bryophyte flora, including mats of *Macrocoma tenuis, Macromitrium urceolatum, Sematophyllum helenicum* and, locally, *Sainthelenia athroclada.* Non-native woodland at middle altitudes is generally poor in bryophytes, and often only *Sematophyllum helenicum* is to be found. However, the longer established woods, such as those around Plantation House, have a more varied flora.

LEFT Cloud forest bryophyte assemblage including *Kurzia nemoides* on a tree fern CENTRE Trunk of Monterey cypress with mats of *Macromitrium urceolatum* RIGHT Peaty bank below tree ferns, Diana's Peak National Park

wet mats or tufts by moderate pressure between the palms of the hands.

Specimens should be air-dried straight away, or at least within 2–3 days of collection to avoid fungal damage. Packets can be hung in paper or mesh bags on a line, or laid out in the sun. Light artificial heat is acceptable, but strong heat should not be used, since 'cooking' will distort cell structures, preventing reconstitution on re-wetting, and rendering the specimen unsuitable for later study. Fibrous absorbent tissue should not be applied to the top of the specimen to aid drying, as this can leave numerous fine fibres all over it, which can present problems if the specimen is studied or photographed later.

The completely dried specimens can be kept permanently in the field packets (so long as they are fully labelled), but fresh herbarium packets are better (Fig. 2). Details should be written or typed on the flap (as shown), not on the back of the packet. This convention serves both for convenience and fits with the practices of international herbaria. The packets can be conveniently stored in shoe boxes in a dry place. In this way, collections can survive in good condition for years (or centuries) providing they are kept dry. However, insect damage is an ever-present threat. If insect pests are detected, then sealing batches of packets in polythene bags and placing them in a freezer for a fortnight should kill any insects they contain.

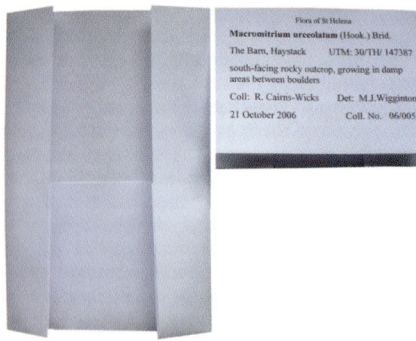

Figure 2. Folding and labelling a herbarium packet

Conservation

St Helena's endemics face a number of threats. In the past, habitat loss has probably been the most serious of these, mainly through the destruction of native forests which are now very scarce and fragmented. The spread of larger invasive plants has also been detrimental: Tall herbs create dense shade and compete for space with black cabbage and other native trees, whose damp bark provides excellent habitat; grasses infest open ground which would once have been filled with bryophyte cushions; and vine-like weeds such as small fuschia displace native epiphytes on trunks and branches. The invasive moss *Pseudoscleropodium purum* has spread rampantly along paths and clearings in the cloud forest zone, out-competing mats of native bryophytes.

Pseudoscleropodium purum invading tussock of *Campylopus arcuatus*, Diana's Peak

Whilst many of St Helena's bryophytes are widespread globally, the endemics have very restricted ranges even within the island, and a few further species (e.g. *Pleurozia gigantea*) are of high conservation importance as they are locally very rare and found only sporadically elsewhere. Four endemics (*Cephalozia sanctae-helenae, Lejeunea sanctae-helenae, Sphagnum helenicum* and *Tylamanthus anisodontus*) appear to be extremely threatened, with only one or two known localities, although careful searching could reveal more. *Philonotis heleniana* and *Sainthelenia athroclada* are almost as scarce. With such tiny populations, these unique members of the island's flora are precariously close to extinction. In 2010, the patch of *Sphagnum helenicum* near Cuckold's Point was severely threatened by an infestation of soft rush, and only a timely intervention by the Peaks Conservation Team saved it.

In order to avoid inflicting further damage on already highly sensitive bryophyte communities, a conservation-minded approach to collecting should always be adopted. The general rule should be to collect as little as possible, so long as there is enough to identify the species. Only part of the bryophyte tuft or mat should be collected, and care should be taken not to dislodge the rest from its substrate. Sometimes just a few stems will be sufficient to confirm a suspected identity, but for an unfamiliar species more will be needed to make an adequate reference specimen. It should also be borne in mind that if a specimen needs to be sent to an expert for identification, then there should be enough material for the range of any variation to be seen, and for a duplicate to be retained. Therefore, circumstances will vary, but with experience collectors will gain an instinct for how much (or how little) should be collected.

Collecting in Diana's Peak National Park should, of course, be minimal, and only the smallest amount should be taken for study. This should apply especially to the native cloud forest communities. There is no need to collect the large, showy and very localised *Pleurozia gigantea*, and some of the species growing epiphytically on the native trees and shrubs are in very small quantities. Also, take care not to spread invasive mosses such as *Pseudoscleropodium purum* on clothing or boots when walking in upland areas.

Microscopic examination

Although many bryophytes can be positively identified with a hand lens, others cannot be certainly identified without microscopic examination. Therefore, those who wish to pursue the study of bryophytes will eventually need the use of a compound microscope with ×4, ×10 and ×40 objectives in order to view slide preparations with transmitted light. It need not be expensive, and either a microscope with a light built into the base, or one with a mirror and separate light (or the mirror angled for daylight) would be perfectly adequate. For distinguishing between species, it is sometimes necessary to compare the relative sizes of various parts of the plant, including leaves and cells. For measurements at the micrometre scale, a graticule (micrometer) for one of the eyepieces will be required, ideally one with a 10 mm scale divided into 100 units.

In addition to a compound microscope, a 'dissecting' stereomicroscope is very useful for viewing specimens up to ×40 magnification, and is ideal for detecting small species that might have been overlooked in the field (this is a frequent experience when examining mixed bryophyte mats). It also makes it much easier to find fertile structures and to dissect off leaves and other parts for examination under high power. Like the compound microscope, a stereomicroscope need not be expensive: for many years, the author used a perfectly satisfactory second-hand one that cost fifty pounds. However, the microscopes should always be designed for serious study, and those sold in hobby shops may not be adequate for studying mosses.

One or two pairs of fine forceps will be needed for removing parts of the plant for examination; extra fine no. 4 (Fig. 3) would be a good choice, though they are not cheap and the tips are easily damaged by careless handling. The illustrated ones are sometimes marketed as 'jewellery tweezers'. Glass slides and square cover slips (no. 1½) are essential, and a water dropper for slide preparation is useful. To examine a leaf, carefully pull it from the stem in a downward direction (making sure the whole leaf comes away from the stem), place it in a small drop of water on the glass slide and cover with a cover slip. If necessary a small amount of water can be added at the edge of the cover slip, or excess removed with a tissue.

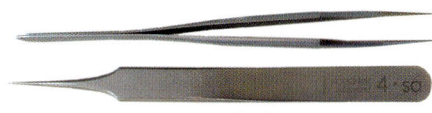

Figure 3. A pair of no. 4 extra fine forceps

Life cycle of bryophytes

The bryophyte life cycle involves an alternation of a **gametophyte** generation which reproduces sexually, and a **sporophyte** generation which reproduces via non-sexual **spores** (Fig. 4). The gametophyte, i.e. the leafy or thallose plant that we recognise in the field as a moss, liverwort or hornwort, produces the sexual cells (antherozoids and eggs) that fuse and germinate into the sporophyte. The sporophyte consists of the **capsule**, **seta** or stalk (except in hornworts) and **foot**. The sporophyte is never free-living and the foot attaches it to the gametophyte throughout its life. The spores are released from the mature capsule, and each develops into a **protonema** (a minute filamentous or filmy structure) from which the new gametophyte grows.

The spores are usually released in damp weather as a cloud of minute, dust-like particles which rise with air currents. Most fall near the parent, but a few may be dispersed for many miles, and are thus extremely efficient at colonising new territory. Bryophytes are generally amongst the first species to appear on the slopes of remote volcanoes after eruptions, and would have been early arrivals on St Helena. Some (e.g. *Tortula* species) produce large, thick-walled spores which can resist desiccation, and only germinate after rains. They complete their life cycle rapidly and the green plants may die-back completely during dry periods.

In fertile gametophytes, the male and female organs (the antheridia and archegonia) are protected by **bracts** (modified leaf-like structures) in leafy liverworts and mosses, and are in cavities or other specialised structures in thallose species. In leafy species, the male and female parts together with their surrounding bracts may be conveniently referred to as an '**inflorescence**'. The **antheridium** is a small, rounded body that develops the **antherozoids** (sperm), the antherozoid moving through water on the plant surface to fertilise the egg at the bottom of the flask-shaped **archegonium**. The resulting fertilised **zygote** eventually develops into the sporophyte. In most bryophytes, the seta is long, raising the capsule well above the leaves or thallus, thus enabling the spores to be efficiently dispersed. However, in a few species (e.g. *Pleuridium acuminatum*) the seta is very short, and most of the capsule remains hidden amongst the leaves at the top of the stem. When capsules are present, bryophytes are commonly referred to as '**fruiting**', even though this term is technically incorrect.

The male and female organs may be borne on separate plants (= **dioicous** species) or in various arrangements on the same plant (= **monoicous** species); within the monoicous group, **autoicous** species are those in which the male and female organs are borne on different branches of the plant.

Non-sexual (vegetative) reproduction is also widespread in bryophytes, though its importance varies greatly between species. In some it may be the most common, or even the only, form of reproduction, and in others it may be unimportant or unknown. New plants can develop from specialised structures including leafy **bulbils** (*Bryum dichotomum*), **tubers** attached to the root-like rhizoids (*Bryum klinggraeffii*) or **gemmae** from leaf tips (*Anastrophyllum subcomplicatum*, *Cephaloziella* spp.). New plants can also regenerate from detached vegetative parts of the plant (branches, leaves, etc). It is entirely possible, though not observed on St Helena, that new plants of, for example *Radula fulvifolia*, may develop from its deciduous, detached leaves.

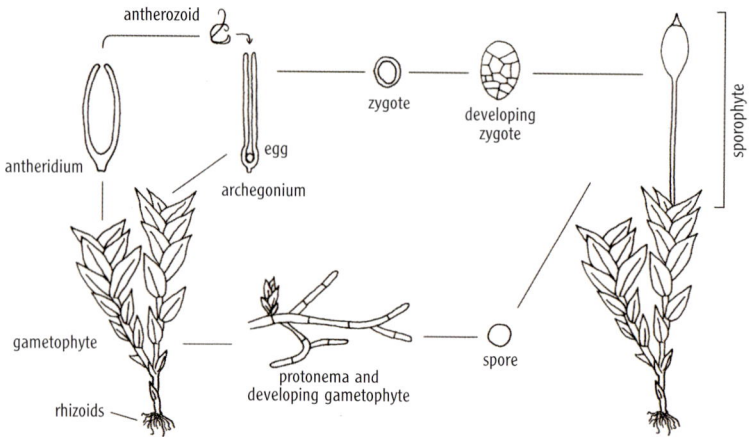

Figure 4. Life cycle of a monoicous (autoicous) moss

Features of liverworts, hornworts and mosses

The three main groups of bryophyte are **liverworts** (hepatics), **hornworts** and **mosses**. The liverworts number about 8,000 species worldwide, the hornworts fewer than 200, and mosses about 12,000. Each group has distinctive or unique characters, the structure of the fertile parts and sporophyte providing the most absolute differences. When identifying species in the field, however, it is often the leafy parts that provide the most immediate guide. Some of the most easily observed features differentiating the three groups are given in a list before the main key to species (p. 13).

LIVERWORTS
The liverworts divide into two natural groups: the leafy liverworts and the thallose (or thalloid) liverworts.

a) Leafy liverworts
Leafy liverworts (Fig. 5) may be prostrate, ascending, erect, or sometimes **pendent** (hanging) from rocks or trees. Erect-growing plants are sometimes supported by a creeping stem, above- or below-ground, anchored by rhizoids.

The leaves provide essential characters for identification, including their shape, structure, and how they are positioned. In almost all leafy liverworts, the leaves are strictly arranged in 2 or 3 rows along the stem. Some leafy liverworts have only the two rows of lateral leaves, but many have an additional row of leaves, the **underleaves**, on the underside of the shoot. The underleaves are almost always smaller than the lateral leaves, and they may sometimes be minute and only detectable with a microscope.

The leaves are attached to the stem in a variety of ways. They may be attached at right angles (**transverse**), or at an **oblique** angle. In some species, the attachment may be different in different parts of the leaf, transverse on one side of the stem and oblique on the other (e.g. in *Anastrophyllum subcomplicatum*). The leaves may be undivided or lobed (sometimes deeply so, as in *Kurzia nemoides*), and always lack a central nerve. They may be flat, concave, convex or channelled, erect to spreading from the stem, and the margin toothed in some species. In many genera, the leaves are divided into two unequal portions, a usually larger **lobe** and a smaller **lobule**, which are folded together. The lobe is on the upper side of the shoot, and the lobule on the underside. The lobule may be attached to the lobe by only a few cells (in *Frullania*) or along a long folded edge or **keel** (in species of the Lejeuneaceae). In *Radula*, it is broadly attached to both the lobe and the stem. All species of the Jungermanniales on St Helena lack a lobule.

The leaf cells of liverworts are almost always about as long as wide, or not much longer than wide. Some have thickenings (**trigones**) of various sorts at the junctions of the cell walls, which may be useful for identifying species, and the cells in some species have projections from their surface (**papillae** or **mammillae**). Glistening, oil-containing structures known as **oil bodies** are found in the cells of most liverworts (but not hornworts or mosses). Their size, number per cell and form provide a valuable aid to identification for genera and species, although a detailed description of these features is outside the scope of this book.

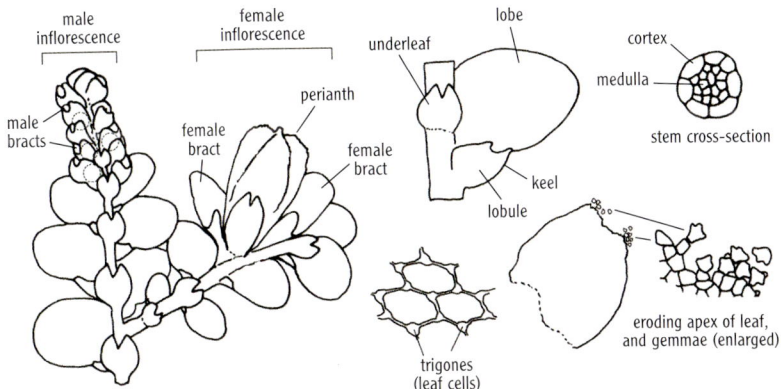

Figure 5. Features of leafy liverworts

plane (e.g. *Lepidopilidium*). Some species have different forms of leaf on the same plant, those along the stem differing from those on the branches (e.g. *Kindbergia praelonga*). The leaves surrounding the archegonia or seta are known as **perichaetial** leaves, and are often markedly larger and/or narrower than those on the rest of the shoot. The leaf **nerve** is an important taxonomic character: it may be long and extend beyond the leaf blade (or lamina) in a hair-point, disappear in the leaf apex, be very short and double or lacking altogether. Leaf shapes are varied, but the leaves of mosses are never divided into lobes. The leaf margins also provide key characters: they may be flat, recurved or incurved, toothed or untoothed, or more than one cell thick. In mosses, the cells along the margins may differ markedly in shape from those in the rest of the blade. Unlike in the leafy liverworts, the leaf cells in mosses are extremely varied in shape, ranging from about as long as wide (e.g. *Tortula muralis*) to very long and narrow (e.g. *Isopterygium*). Their walls may be thin or thickened, sometimes unevenly so.

Capsules usually develop on an elongate seta, though the seta is very short in some genera. The shape, size, structure and orientation of the capsule is sometimes important in identifying species, and some species on St Helena (e.g. *Weissia* species) can only be separated on characters of the capsule (and spores). Moss capsules are varied in shape, from almost spherical to long cylindrical, but the most common form is shortly cylindrical. They may be upright or variously inclined or pendent. The **urn** (body of the capsule) is closed by an **operculum** (lid), and in most species, the operculum and part of the urn is covered by a thin **calyptra** that protects the maturing capsule. The operculum is variable in form, and may be flat (*Entosthodon* species) to conical or long-beaked (rostrate) (*Dicranella proscripta*). The calyptra falls first, then the operculum becomes detached at maturity, allowing the spores to escape. The capsules of most mosses have a **peristome** that consists of one or two rows of variously ornamented teeth around the mouth. They are revealed when the operculum becomes detached, opening and closing according to atmospheric humidity and thus controlling the release of the spores. The number, shape and ornamentation of the peristome teeth can be useful in distinguishing genera or species: they can be short and blunt to long and narrow. Some species (e.g. *Entosthodon* sp. and *Weissia brachycarpa* in the St Helena flora) lack a peristome, and others have only a short or rudimentary one.

Exceptionally large, almost 'pure' stand of *Kurzia nemoides*

TOP **Leafy liverwort (left); thallose liverwort (centre); hornwort (right)**
BOTTOM **Acrocarpous moss (left); pleurocarpous moss (right)**

How to recognise the three main groups of bryophytes

Each of the three main groups of bryophyte has unique attributes that can be used to determine to which group a specimen belongs. The most absolute of these are found in the sporophytes, but there are also vegetative characters unique to, or at least highly indicative of, a particular group. For example, leaf nerves are found only in mosses (but leaves lacking a nerve may belong to a moss or a liverwort). Other features, though not unique, are almost exclusively or predominately found in only one group. For example, almost all bryophytes with 3 rows of leaves are liverworts. Conversely, leaves that have a distinct border of cells that differ from the other laminal cells will almost always belong to a moss.

Some of the most easily observed features of the three groups are given below, but you will quickly come to recognise at a glance into which category a plant should be placed.

Plant thallose, not differentiated into stem and leaves .... liverwort or hornwort
Plant with underleaves (a row of leaves on the underside of the plant that are usually
    smaller than the other leaves, and differing in shape) .... liverwort
Plant with leaves spirally arranged (sometimes the shoot is flattened and leaves apparently
    in several rows) .... moss
Leaf divided into lobes .... liverwort
Leaf with a nerve, sometimes very short and forked (some thallose liverworts have a midrib) .... moss
Leaf with a distinct border of cells different from those in rest of the leaf blade .... moss
Leaf cells long and narrow, or cells papillose .... moss
Capsule globular or ellipsoid, on a translucent, fragile, short-lived, pale seta (stalk) .... liverwort
Capsule on a firm, coloured, persistent seta .... moss
Rhizoids unicellular, usually whitish or reddish .... liverwort
Rhizoids multicellular, usually brownish .... moss

Keys to the identification of species or genera

Several species, such as *Bryum argenteum, Pleurozia gigantea* and *Pseudoscleropodium purum*, are so distinctive that they can be recognised from some distance away. Other species can, with experience, be confidently identified in the field with a hand lens, and features that can be seen with a hand lens have been favoured in the following keys. However, many species cannot be positively identified without microscopic examination, and for this reason, microscopic characters are included in the keys and species accounts.

These keys are designed to help you quickly find the right match for your specimens, and in most instances you should be able to do so without much trouble. They do include a fair number of technical terms, which may be a stumbling block at first, but definitions and illustrations are provided in the glossary and you will quickly become familiar with these special terms. One drawback of keys is that they include only a limited number of features of the plant, and cannot always encompass all the variability that a species might show. Thus, following the keys might occasionally lead to the wrong conclusion. It is therefore essential that all keyed identifications are confirmed by carefully checking the species descriptions and illustrations.

Observers should also be alert to the fact that the St Helena bryophyte flora has not been fully investigated, and that if a plant does not seem to match a species account and illustration, then it might prove to be new to the island. Stunted or otherwise atypical plants not displaying features typical of the species may also cause confusion from time to time, as also might young plants. Some of these forms will defy identification, so do not expect to be able to name everything, even if the material appears to be in good condition. However, it is worth keeping a sample of the plants you cannot identify, in case they can be identified later on.

Using the keys

The keys are arranged in pairs throughout. Choose which sentence of the pair better fits your plant, then follow the number to the next pair of choices. For example, in the pair (1a–1b) at the start of the key: if your plant does not have leaves (i.e. it is a thallus), go to 2, then make a choice between the two options, and follow the option that matches your specimen, and so on down the key; if your plant does have leaves, then go to 17 in the key, and work down the key from there. The key is followed through in this way until you arrive at a species name. The result should be checked by referring to the descriptions and photographs.

Key to the liverworts, hornworts and mosses of St Helena

1   a) Plants without leaves, in the form of a thallus 2
    b) Plants with leaves (mosses and leafy liverworts) 17
2   a) Thallus one cell thick throughout. A disc-like or strap-like body, unlike any
       thallose liverwort described in this guide (see also prothallus entry in the glossary) .... a fern prothallus
    b) Thallus more than one cell thick, at least in the middle or in a midrib
       (hornwort or thallose liverwort*) 3
*A few lichens (especially *Collema* and *Leptogium* spp.) have a soft, thallus-like growth form, but are generally gelatinous (jelly-like) when wet, crust-like when dry, and are often coloured shades of blackish, brown or grey.

Hornworts and thallose liverworts

3   a) Thallus with an elongate, horn- or stick-like capsule 4
    b) Thallus lacking an elongate horn- or stick-like capsule 8
4   a) Plants growing directly on stems, branches or twigs of trees and shrubs ...... ***Dendroceros adglutinatus*** (p. 61)
    b) Plants growing on soil, occasionally rock (rarely soil on tree bases) 5
5   a) Plants with mature sporophytes, some darkening and/or splitting into
       two parts at the apex and releasing spores 6
    b) Plants with only immature sporophytes, lacking any with dark tips and
       mature spores not present 7
6   a) Spores yellow. Common plant of shaded moist soil in a range of habitats ...... ***Phaeoceros carolinianus*** (p. 60)
    b) Spores black. Apparently rare, but may be locally frequent in gardens
       or cultivations .... ***Anthoceros*** sp. (p. 59)
7   a) Thallus rather flat, the margins not strongly crisped or convoluted; thallus
       not forming a crinkled cup-, trumpet- or bell-like structure .... probably ***Phaeoceros carolinianus*** (p. 60)
    b) Thallus not flat, the margins crisped or convoluted and ascending to
       form a crinkled cup- or trumpet- or bell-like structure .... probably ***Anthoceros*** sp. (p. 59)

8 a) Plants growing directly on stems, branches or twigs of trees and shrubs 9
   b) Plants growing on soil or rock, rarely soil on tree bases 10
9 a) Thallus repeatedly forked (i.e. divided in a Y-shaped fashion into two
   equally-sized branches, which themselves are forked) with a broad
   central midrib .... *Dendroceros adglutinatus* (p. 61)
   b) Thallus regularly or irregularly pinnately branched, lacking a
   well-defined midrib .... *Riccardia* spp. (p. 58)
10 a) Thallus whitish or pale grey, ~3 mm wide, simple or branched, forming
   complete or partial rosettes, the upper surface covered with conical
   protuberances with a pore at the apex .... *Exormotheca pustulosa* (p. 53)
   b) Thallus neither whitish nor pale grey 11
11 a) Thallus less than 1.5 mm wide, regularly or irregularly 1 to 2-pinnately branched ..... *Riccardia* spp. (p. 58)
   b) Thallus usually more than 1.5 mm wide, not pinnately branched 12
12 a) Thallus almost erect or sub-erect, often in crowded tufts, divided into
   2 to several branches, margins of branches with teeth (which may be
   well-spaced and small, or prominent and spine-like) .... *Jensenia spinosa* (p. 56)
   b) Thallus prostrate, margins without teeth 13
13 a) Thallus glaucous-green, irregularly bordered with purple, lacking obvious
   pores on upper surface; female receptacles dome-shaped, raised on
   a short stalk .... *Plagiochasma rupestre* (p. 54)
   b) Thallus mid-green, not purple-bordered 14
14 a) Thallus with obvious pores on upper surface (appearing as whitish dots);
   underside of thallus with scales; male and female receptacles lobed,
   on long stalks .... *Marchantia berteroana* (p. 55)
   b) Thallus without pores on upper surface; underside of thallus without scales 15
15 a) Thallus strap-like, with a broad, often coloured, midrib, margins smooth;
   thallus solid, with a strand of small cells in the midrib (cross-section) .... *Symphyogyna brasiliensis* (p. 57)
   b) Thallus not strap-like, lacking a midrib, margins uneven, crisped or
   convoluted; thallus without a central strand of small cells 16
16 a) Thallus rather flat, the margins not strongly crisped or convoluted;
   thallus solid, lacking internal cavities .... probably *Phaeoceros carolinianus* (p. 60)
   b) Thallus not flat, the margins crisped or convoluted; thallus with large
   internal cavities (cross-section) .... probably *Anthoceros* sp. (p. 59)

Leafy liverworts
17 a) Leaves strictly arranged in 2 rows, or in 3 rows (i.e. one row of underleaves
   and 2 rows of other larger leaves). (*Fissidens*, and leafy liverworts) 18
   b) Leaves not strictly arranged in 2 or 3 rows (mosses, except *Fissidens*) 51
18 a) Leaves without a nerve (leafy liverworts) 19
   b) Leaves with a nerve; leaves oblong-ovate to oblong-lanceolate (*Fissidens* spp.) 46
19 a) Leaves deeply dissected into 3–4 narrow segments; underleaves deeply
   divided into 3–4 segments .... *Kurzia nemoides* (p. 41) and *Lepidozia africana* (p. 42)
   b) Leaves not divided into 3–4 narrow segments 20
20 a) Leaves in the form of a swollen, ovate or cylindrical sac with a narrow
   'beak' (very small epiphytic plants) .... *Colura calyptrifolia* (p. 33) and *C. tenuicornis* (p. 34)
   b) Leaves not modified into a swollen sac with a narrow 'beak' 21
21 a) Leaves divided into a lobe and a smaller lobule 22
   b) Leaves not divided into a lobe and a lobule 35
22 a) Plants very robust, usually deep red or sometimes reddish-yellow; leaves
   divided into a strongly convex, folded lobe and much smaller horn-
   shaped lobule .... *Pleurozia gigantea* (p. 22)
   b) Plants smaller, not as above 23
23 a) Lobules broadly attached to the stem; leaf lobes deciduous, so that often
   only the much smaller lobules remain attached to the stem .... *Radula fulvifolia* (p. 23)
   b) Lobules narrowly attached to the stem 24

24 a) Lobules only partially attached to the lobe; lobules small, hood- or
helmet-like; underleaves present; plants olive-green to blackish .... *Frullania depressa* (p. 24)
  b) Lobules attached for most of their length to the lobe along a keel;
lobules not hood- or helmet-like (except in *Lejeunea sanctae-helenae*);
underleaves present or absent (Lejeuneaceae) 25
25 a) Underleaves absent 26
  b) Underleaves present 29
26 a) Plants minute and delicate, thread-like, less than 0.5 mm wide; leaves less than 0.3 mm long ....
.... *Cololejeunea microscopica* (p. 30) and *C. minutissima* (p. 31)
  b) Plants small, more than 1 mm wide; leaves more than 0.5 mm long 27
27 a) Leaves pointed at apex .... *Cololejeunea sanctae-helenae* (p. 32)
  b) Leaves broadly rounded at apex 28
28 a) Shoots 1.5–2.8 mm wide; stylus present at base of lobule; perianths
with 5 long, sharp, narrow keels .... *Cololejeunea grossestyla* (p. 29)
  b) Shoots 1–1.3 mm wide; stylus absent; perianths with 5 prominent
spreading horns at the apex .... *Cololejeunea dianae* (p. 28)
29 a) Underleaves rounded at apex, not incised or divided into lobes 30
  b) Underleaves divided into lobes 32
30 a) Plants deep olive- to blackish-green; lobules with 2–4 small, prominent
teeth; perianths strongly flattened, lacking a keel on both surfaces .... *Marchesinia brachiata* (p. 38)
  b) Plants pale or whitish-green, olive or brownish; lobules with 1 blunt tooth
or almost untoothed; perianths with a large, broad keel on upper or lower surface 31
31 a) Underleaves large, wider than long (often kidney-shaped); lobes with
strongly inrolled apex and margins; plants often brownish, shoots
compressed-cylindrical (worm-like) .... *Aureolejeunea microscypha* (p. 25)
  b) Underleaves smaller, rounded, about as wide as long; lobes convex,
apex and margin flat; plants pale green, shoots not worm-like .... *Aureolejeunea rotalis* (p. 26)
32 a) Leaf lobe apex with a short, sharp point (but apex sometimes turned
down in the moist leaf, so that the point may not be immediately
obvious in the field), or sometimes lobe bluntly pointed .... *Cheilolejeunea ascensionis* (p. 27)
  b) Leaf lobe apex broadly rounded 33
33 a) Plants medium-sized, shoots 1.3–2 mm wide; lobules typically
very small .... *Lejeunea eckloniana* (p. 36)
  b) Plants small, shoots up to 0.6 mm wide; lobules larger, sometimes more
than half the length of the lobe 34
34 a) Plants very small, or delicate and thread-like, shoots 0.2–0.35 mm wide;
leaves almost parallel to stem; lobules proportionately large, often more
than half the length of the lobe .... *Microlejeunea africana* (p. 39)
  b) Plants larger, shoots 0.4–0.6 mm wide; lobule smaller, usually $1/3$ lobe
length; leaves spreading from stem .... *Lejeunea autoica* (p. 35)
35 a) Underleaves present 36
  b) Underleaves absent (or minute and not detectable in the field) 39
36 a) Leaf apex lobed or toothed 37
  b) Leaf apex rounded, truncate, or slightly notched 38
37 a) Leaf apex deeply divided into two long, pointed lobes; underleaves
deeply divided into 2 toothed lobes ....*Chiloscyphus coadunatus* (p. 43)
  b) Leaf apex with 2–3 short, broad teeth; underleaves not divided into lobes;
some branches arising at a wide angle from underside of stem .... *Bazzania praerupta* (p. 40)
38 a) Underleaves divided into 2 triangular lobes; plants usually pale green ..... *Chiloscyphus humistratus* (p. 44)
  b) Underleaves rounded, not divided into lobes; plants usually brownish
(sometimes green) .... *Mnioloma fuscum* (p. 52)
39 a) Leaves deeply divided into 2 almost equal lobes 40
  b) Leaves toothed, or divided into 2 unequal lobes 44
40 a) Leaf lobes blunt, sometimes converging .... *Cephalozia sanctae-helenae* (p. 47)
  b) Leaf lobes acute (but may be eroded by gemmae in *Anastrophyllum subcomplicatum*) 41

41 a) Leaves transversely inserted on stem; plants minute, thread-like .... *Cephaloziella* spp. (p. 48)
   b) Leaves obliquely or nearly longitudinally inserted on stem 42
42 a) Leaves strongly concave, acute lobes sometimes not obvious in the field
   (they may often be eroded, the leaf appearing irregularly rounded at apex);
   plants reddish-brown to deep magenta or dark purple-black .... *Anastrophyllum subcomplicatum* (p. 50)
   b) Leaves weakly concave to almost flat; plants pale to dark green 43
43 a) Plants minute; leaves widest at base, divided into acute but not spinose
   lobes; cylindrical perianths often present .... *Cylindrocolea helenae* (p. 49)
   b) Plants larger, stem often partly denuded of leaves; leaves narrowed to the
   base (widest near the middle), divided into 2 almost spinose lobes .... *Plagiochila spinulosa* (p. 45)
44 a) Leaves rounded with 2 small, spinose teeth at apex, the front margin
   usually narrowly incurved; plants usually dark-coloured .... *Adelanthus decipiens* (p. 46)
   b) Leaves oblong-rounded; plants green 45
45 a) Leaf apex and back margin with 4–8 spinose teeth, front margin reflexed;
   leaves often deciduous, and stems often partly denuded of leaves .... *Plagiochila spinulosa* (p. 45)
   b) Leaf apex unequally 2-lobed, or irregularly toothed; margins not toothed;
   leaves not deciduous, stems not denuded of leaves .... *Tylimanthus anisodontus* (p. 51)

Fissidens (see glossary figure for an explanatory drawing of a *Fissidens* leaf p. 123)
46 a) Plants medium-sized to large, shoots to 10 mm or more long 47
   b) Plants small to minute, shoots 1–8 mm long 48
47 a) Leaves ovate-lanceolate, apex obtuse, nerve excurrent in a rather long point,
   leaf margin minutely serrate .... *F. taxifolius* (p. 74)
   b) Leaves lanceolate, gradually narrowed to an acute apex; nerve ending in apex,
   leaf margin not serrate .... *F. chioneurus* (p. 66) and *F. elegans* (p. 69)
48 a) Leaves with a conspicuous border (sometimes discontinuous) of translucent narrow cells
   in all laminae (all margins are bordered in the smaller *F. curvatus* subsp. *helenuicus*,
   but they are often difficult to observe in the field with a hand lens) .... *F. reimersii* (p. 72)
   b) Leaves lacking a conspicuous border 49
49 a) Leaf nerve sharply angled in middle of leaf, so the upper part of the leaf is
   bent to one side (but nerve nearly straight in upper leaves of fertile shoots);
   leaves broadly ovate-lanceolate .... *F. pygmaeus* (p. 71)
   b) Leaf nerve straight 50
50 a) Leaves tongue-shaped, apex blunt, margins crenulate, sheathing laminae
   with border of translucent cells (seen with difficulty in the field) .... *F. darntyi* (p. 68)
   b) Leaves lanceolate or narrowly ovate-lanceolate, margins various, laminae bordered
   or not (microscope required for the identification of these small to minute species) ....
   .... *F. curvatus* subsp. *helenicus, F. porrectus, F. serratus, F. taylorii, F. tenellus, F. translucens*
   (p. 67–76)

Other acrocarpous mosses
51 a) Plants with short, hanging branches spirally arranged along stem, the branches
   compacted to form a head (capitulum) at stem apex; leaves composed of a
   network of narrow green cells surrounding large, colourless cells .... *Sphagnum helenicum* (p. 62)
   b) Plants not as above 52
52 a) Stems/shoots erect, solitary or forming tufts or cushions; sporophytes arising
   from the tips of main shoots (acrocarpous species) 53
   b) Stems/shoots creeping, spreading or pendent, pinnate or irregularly
   branched, sometimes with abundant, short, erect branches; sporophytes not
   arising from the tips of main shoots (pleurocarpous species) 80
53 a) Upper part of leaf silvery-white (tuft of shoots appearing all silvery-white);
   leaves broad, strongly concave, tapering to a long, acute apex .... *Bryum argenteum* (p. 96)
   b) Upper part of leaf not silvery-white (whole tuft not appearing silvery-white) 54
54 a) Leaf ending in a hyaline hair-point, which appears whitish and especially
   conspicuous in the dry plant 55
   b) Leaf not ending in a hyaline hair-point 56

55 a) Leaves narrowly lanceolate, gradually tapering to an acute apex .... *Campylopus introflexus* (p. 81) and *C. pilifer* (p. 82)

b) Leaves tongue-shaped, quickly tapering to an obtuse apex .... *Tortula muralis* (p. 91) and *Pseudocrossidium crinitum* (p. 89)

56 a) Basal part of leaf mostly composed of very large, transparent cells (looking like large clear panels on each side of the nerve) strongly differentiated from the small green cells of margins and upper part of leaf; leaves strongly crisped when dry; plants often yellowish- or whitish-green, forming tufts .... *Syrrhopodon gaudichaudii* (p. 85)

b) Basal part of leaf not composed mostly of very large, transparent cells 57

57 a) Plants in tufts on wet rocks by streams and waterfalls; shoots green with paler tips; leaves triangular-lanceolate, usually curved to one side .... *Philonotis heleniana* (p. 103)

b) Plants of moist or dry habitats, but not on constantly wet rocks by streams and waterfalls 58

58 a) Leaves narrowly lanceolate, with a border of extremely narrow, long, thick-walled cells strongly contrasting with the elongate-hexagonal cells in the rest of the leaf; plants rare, forming small tufts on trunks and branches .... *Daltonia splachnoides* (p. 108)

b) Leaves unbordered, or if bordered, then marginal cells not extremely narrow, long and thick-walled 59

59 a) Leaves narrowly lanceolate or narrowly triangular, 6 times as long as wide or more, and gradually or quickly narrowing to a long, acute point 60

b) Leaves ovate, lanceolate or tongue-shaped, usually 2–5 times as long as wide, or if narrower, then not tapering to a long, acute point 67

60 a) Upper leaves rapidly narrowing from a broad, almost rectangular base to a very slender limb that tapers to a narrow point 61

b) Upper leaves not rapidly narrowing from a broad base to a slender limb (leaf shapes various) 62

61 a) Shoots usually 2–4 cm high; upper leaves strongly spreading-flexuose; seta 6–10 mm long, flexuose, capsule not downwardly-directed .... *Dicranella proscripta* (p. 83)

b) Shoots usually 0.5–2 cm high; upper leaves erect, spreading or all curved to one side; seta 4–6 mm long, flexuose, or sometimes strongly curved and the capsule downwardly directed .... *Dicranella* spp. (p. 84)

62 a) Upper and perichaetial leaves much longer than lower leaves; leaves linear or narrowly lanceolate, gradually tapering to a long, fine point 63

b) Upper and perichaetial leaves not much longer than lower leaves 64

63 a) Upper leaves flexuose, widely spreading; stem below with long brown rhizoids often bearing brownish tubers; capsules (not observed on St Helena) borne on a long seta. Frequent in flower pots .... *Leptobryum pyriforme* (p. 95)

b) Upper and perichaetial leaves not flexuose, almost erect; capsules on a very short seta, immersed in the perichaetial leaves; stem without dark rhizoids, tubers absent .... *Pleuridium acuminatum* (p. 78)

64 a) Leaves narrowly triangular, sharply toothed from base to apex, the nerve narrow at leaf base; shoots with matted, reddish or brownish tomentum below; plants forming pale or yellowish-green tufts .... *Philonotis helenica* (p. 104)

b) Leaves not sharply toothed from base to apex, the nerve often wide at leaf base; shoots with or without tomentum 65

65 a) Leaves not strongly contorted, twisted, curled or crisped when dry 66

b) Leaves strongly contorted, twisted, curled or crisped when dry, spreading-erect when moist; plants forming small, compact tufts or mats 67

66 a) Cells in mid-leaf narrowly rectangular; plants small, usually 0.5–2 cm high, forming dense mats .... *Dicranella* spp. (p. 84)

b) Cells in mid-leaf short (oval or rhomboidal); plants usually more robust, up to 7 cm high .... *Campylopus* spp. (p. 79–82)

67 a) Leaves rather tightly curled or crisped when dry, spreading-erect when moist; leaves tongue-shaped or narrowly lanceolate, apex obtuse or acute; plants forming small, compact tufts or mats 68

b) Leaves contorted, twisted or shrunken when dry, but not tightly curled or crisped; leaves lanceolate or ovate, apex not obtuse; plants in compact tufts or not 70

68 a) Leaf apex hooded (shaped like the prow of a boat) .... *Trichostomum crispulum* (p. 93)
    b) Leaf apex not or only very slightly hooded (see also description of *Weissia* spp.) 69
69 a) Leaves narrowly lanceolate; leaf margins usually rolled inwards above .... *Weissia* spp. (p. 94)
    b) Leaves wider, often tongue-shaped; leaf margins flat to erect above ....
    .... *Chionoloma bombayensis* (p. 86) and *Trichostomum brachydontium* (p. 92)
70 a) Capsules pear-shaped, broadly ovoid or sub-globose, erect, on a long seta 71
    b) Capsules ovoid or cylindrical, erect, inclined or pendent, or immersed in
    the perichaetial leaves, or capsules absent 73
71 a) Leafy shoots elongate, 2–3.5 cm high; leaves not crowded at apex;
    capsules sub-globose, lid almost flat, with a short prominent rostrum ..... *Physcomitrium flexifolium* (p. 64)
    b) Leafy shoots shorter, 0.5–2 cm high; leaves crowded at apex; capsules
    pear-shaped or ovoid; lid various (flat, conical, or with a short, prominent rostrum) 72
72 a) Capsule broadly and markedly pear-shaped, gradually tapering at the base;
    lid small, almost flat, lacking a rostrum .... *Entosthodon* sp. (p. 63)
    b) Capsule broadly ovoid or slightly pear-shaped, more abruptly tapering at
    the base; lid bluntly conical or shortly rostrate .... *Physcomitrium* sp. (p. 65)
73 a) Plants small, forming dark green or blackish tufts or mats; leaves spirally-
    twisted when dry, nerve very broad and thickened in upper half of leaf .... *Tortula atrovirens* (p. 90)
    b) Plants not dark or blackish; leaf nerve not thickened in upper part of leaf 74
74 a) Leaves broad, widest above mid-leaf, widely spreading from the stem when moist, with
    reflexed, acute apices; dry leaves curved, often appearing claw-like, with folded or incurved
    margins; plants rather soft, the dry leaves often deciduous .... *Leptophascum leptophyllum* (p. 88)
    b) Leaves ovate, ovate-lanceolate or lanceolate; dry leaves not appearing claw-like, not deciduous 75
75 a) Plants robust; upper leaves large, 2–3.2 mm long, sometimes crowded at the
    top of the shoot; dry leaves appressed, shrunken, the apical point reflexed .... *Bryum canariense* (p. 97)
    b) Plants smaller; leaves not crowded at the top of the shoot 76
76 a) Leaf margins narrowly recurved only in the lower part of the leaf; leaf cells
    narrowly rhomboidal-hexagonal, 4–6 times as long as wide 77
    b) Leaf margins narrowly recurved from near leaf base to near apex; leaf cells
    square or rounded, about as wide as long 79
77 a) Leaves broadly ovate, concave; yellowish-green bulbils in leaf axils (some plants
    in a tuft may lack bulbils) .... *Bryum dichotomum* (p. 98)
    b) Leaves rather narrower, not markedly concave; bulbils absent; rounded or
    pyriform tubers developed on coloured rhizoids (*Bryum erythrocarpum* group) 78
78 a) Red or crimson spherical tubers clustered around base of stem (×20 hand lens) ...... *Bryum rubens* (p. 100)
    b) Tubers, if present, on long usually subterranean rhizoids (soil should be carefully
    washed from the plant, ensuring that tubers remain attached to the rhizoids) ....
    .... *Bryum klinggraeffii, B. radiculosum, B. rubens, B. sauteri, B. subapiculatum* (p. 99–102)
79 a) Shoots very small, 2–3 mm high; cells in mid leaf papillose (one papilla per cell);
    leaf margins crenulate .... *Didymodon* sp. (p. 87)
    b) Shoots larger, 1–3.5 cm high; leaf cells smooth; leaf margins smooth .... *Ceratodon purpureus* (p. 77)

Pleurocarpous mosses
80 a) Plants with abundant, crowded, ascending-erect branches; archegonia (therefore
    also capsules) borne at the apex of branches (cladocarpous species); plants forming
    dense mats on trees and, less frequently, on rocks 81
    b) Plants prostrate, pinnately or irregularly branched, branches not erect-ascending;
    archegonia (and capsules) from small buds on stems or branches 83
81 a) Leaves closely appressed-erect when dry .... *Macrocoma tenuis* (p. 105)
    b) Leaves strongly crisped and curled when dry 82
82 a) Leaf apices fragile; most leaves with broken, truncate tips .... *Macromitrium urceolatum* (p. 107)
    b) Leaf apices not fragile; leaves with intact, acute tips .... *Macromitrium microstomum* (p. 106)
83 a) Leaves strongly curved to one side or downwards; leaves ovate or ovate-lanceolate 84
    b) Leaves not curved to one side or downwards, or if slightly curved downwards,
    then shoots flattened in one plane 86

84 a) Shoots regularly pinnate, pale green .... *Hypnum jutlandicum* (p. 116)

b) Shoots more irregularly branched; mid-green or golden brownish-green 85

85 a) Shoots cylindrical (rather swollen-looking) when moist, plants usually
golden brownish-green .... *Hypnum lacunosum* (p. 117)

b) Shoots more flattened when moist, plants green .... *Hypnum cupressiforme* (p. 115)

86 a) Leaves with a nerve to mid leaf or beyond 87

b) Leaves lacking a nerve, or nerve obscure 90

87 a) Leaves large, broadly rounded at apex but abruptly terminated by a short,
sharp point; plants robust, branching pinnate; branches blunt,
swollen in appearance .... *Pseudoscleropodium purum* (p. 113)

b) Leaves tapering to an acute apex; branching various, but branches not blunt
and swollen in appearance 88

88 a) Shoots regularly bipinnate; stem leaves broadly triangular, decurrent at the
base, and sharply narrowed above to a long point; branch leaves much
narrower, lanceolate .... *Kindbergia praelonga* (p. 111)

b) Shoots pinnate or not; leaves not as above 89

89 a) Plants small, pinnate, the branches short, densely packed on the stem;
leaves small, narrowly triangular-lanceolate, margins not toothed .... *Sainthelenia athroclada* (p. 114)

b) Plants larger, branching irregular, more widely-spaced on stem; leaves
larger, broadly ovate, the margins sharply toothed throughout .... *Oxyrrhynchium hians* (p. 112)

90 a) Shoots strongly flattened in one plane 91

b) Shoots not strongly flattened in one plane 92

91 a) Plants slender; leaves rather widely spaced on the stem, up to 1.2 mm long,
strongly asymmetric, apex sharply toothed; leaf cells very narrow (5 μm wide)
throughout leaf .... *Isopterygium* sp. (p. 119)

b) Plants more robust; leaves crowded on the stem, to 2 mm long; leaf cells wider
(8–12 μm), becoming shorter and elongate-hexagonal near apex ..... *Lepidopilidium pallidifolium* (p. 110)

92 a) Leaves narrowly lanceolate; leaf bases with group of very large alar cells ..... *Sematophyllum* spp. (120–121)

b) Leaves ovate or ovate-lanceolate; leaf bases with small alar cells, or
alar cells similar to laminal cells 93

93 a) Leaves strongly shrunken and crisped when dry; leaves on upper surface
of shoot raised from the stem when moist, often giving the shoot a
rather spiky appearance .... *Lepidopilidium crispifolium* (p. 109)

b) Leaves not much altered when dry, appressed in the moist shoot .... *Entodon dregeanus* (p. 118)

Exormotheca pustulosa, heavily fruiting

The species accounts

The accounts are arranged taxonomically, i.e. reflecting the species evolutionary tree, according to the best available knowledge. The 'order' (a broad group of related species) is indicated in the figure legend, and the 'family' (a narrower group of closer relatives) is indicated next to the species name. Descriptions of the species include only those characters that have been observed on St Helena specimens, and in a few cases this may not fully cover the variation seen throughout its worldwide range. The descriptions are sometimes more detailed than might be considered usual for a field guide, including, for example, the sizes of cells and characters of the tubers of *Bryum* species. However, it was considered desirable to provide such additional information because there is no detailed bryophyte flora for the island; such information is very dispersed in the literature, and often not readily available. This book may, therefore, be regarded as a 'halfway house' between a basic field guide and a technical flora.

In the species accounts, the sizes given for the various parts of the plant, including leaves, cells and capsules, are those found in typical plants. However, it should be noted that some species can be variable, and that sizes in stunted or atypical plants may lie outside the quoted size or range. Young plants can also be a source of confusion.

Only common names are given for flowering plants, conifers and ferns in the accounts of the species, but their scientific names are provided in the Appendix (p. 126).

The synonyms listed are the names of the species as used in the two early accounts of St Helena bryophytes, namely in Hooker & Taylor (1845) and Mitten (1875), and a few others that have been widely used in more recent years.

All the photographs are of St Helena specimens, and all were taken by the author, except for the following: *Marchesinia brachiata* (inset), *Plagiochasma rupestre* (top right in the colour figure) and *Sphagnum helenicum* by Rebecca Cairns-Wicks; *Macromitrium urceolatum* calyptrae by Andrew Darlow; *Exormotheca pustulosa* sporophytes, *Physcomitrium* sp. and *Weissia* sporophytes by Phil Lambdon. Many species were photographed in situ during field surveys, but for smaller species in particular, coverage was completed using herbarium specimens both dry and wetted.

The species distribution maps have been compiled from data obtained during the 2005 surveys, together with a small amount of data added later. The maps are inevitably incomplete since many areas, especially outside the national park, have not been thoroughly explored, and many areas not visited at all. Nevertheless, they give a useful indication of the geographical distribution of the species, and show where gaps can be filled.

Note that the three summits of Diana's Peak Ridge are here named in accordance with the Ordinance Survey map of the island: Mt Actaeon, Diana's Peak and Cuckold's Peak, named from north to south. Many islanders use the names in the reverse order, and at present the signage on the peaks also reflects this interpretation.

LEFT **Tuft on rock** RIGHT **Shoots**

Pleurozia gigantea

Pleuroziaceae

Synonym *Physiotium sphagnoides*

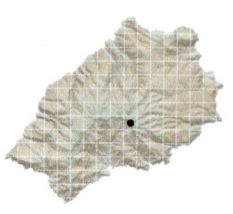

Description Plants deep wine-red to pale red or yellowish, forming tufts; shoots large and robust, erect to somewhat pendent, to 10 cm long and 1 cm wide, sparsely branched. Leaves in two opposite rows, each composed of a large, clasping, concave lobe, and a much smaller, tubular, sac-like lobule attached to the basal half of the lobe. Underleaves absent. Autoicous. Female inflorescences on very short lateral branches hidden amongst the leaves, the female bracts covering the lower part of the perianths; perianths cylindrical, tapered, the fertile ones deeply furrowed, the sterile ones smooth.

Recognition *P. gigantea* is unmistakable, and immediately recognisable from some distance away. The deep red, or at least red-tinged shoots, and the two rows of bilobed, clasping leaves readily identify this species in the field, making it unnecessary to collect specimens for microscopic examination.

Habitats In native habitats, *P. gigantea* now appears to be confined to the branches and trunks of a few black cabbage trees, and a few low rock outcrops. It is frequent on two large Norfolk Island pine trees on Mt Actaeon and Cuckold's Point, especially on high branches, and was recorded from tree ferns in the 19th century. Although not recorded on the latter host since then, further survey may show that tree ferns still support colonies of this liverwort.

Status and distribution It is restricted to the upper altitudes of Diana's Peak National Park, where it occurs in only a few locations. Unfortunately, *P. gigantea* is such a large and attractive plant that its accessible populations are under threat from casual collecting, and it may have become rarer on the island for that reason. *P. gigantea* and *Macromitrium urceolatum* were the first bryophytes to be recorded from St Helena, collected by Archibald Menzies in 1795. *Range*: Widespread but very local in Africa, southern and eastern Asia, and the Pacific Ocean region.

LEFT **Intact shoots** CENTRE **Deciduous shoots** RIGHT **Underside of shoot**

Radula fulvifolia

Radulaceae

Synonym *Jungermannia fulvifolia*

Description Plants pale or mid- to bright green, soft in texture, creeping, forming dense, thin mats or 'sheets'. Leafy shoots 1.5–2 mm wide, 1–2 cm long; branches frequent, arising immediately below a leaf. Leaves closely overlapping, often deciduous, 0.7–1.0 mm long, margins usually crenulate; lobules rhomboidal, broadly attached to, and flattened against the lobe and stem, the apex rounded or bluntly pointed; with a rather sharp angle between the lobule and the lower margin of the lobe. Underleaves absent. Perianths long-cylindrical, trumpet-shaped, 2 mm long, but not yet reported from St Helena.

Recognition *R. fulvifolia* may initially be mistaken for a species of the Lejeuneaceae, especially when in prostrate mats amongst other bryophyte species. However, it is readily recognised by the small lobules widely attached to the stem, and by the lack of underleaves. The deciduous leaf lobes, often resulting in shoots denuded below with only the small lobules remaining on the stem, is a typical feature of this species, and a good field character.

Habitats Rock, soil and trees in shaded locations, in small discrete mats, or in extensive overlapping 'sheets' on steep or vertical substrates. On rock outcrops and cliffs, in crevices, on earthy banks, on tree trunks, often creeping over other bryophytes.

Status and distribution Widespread in the central part of the island, including in Diana's Peak National Park; also on the Barn. *Range*: Tropical Africa, including Madagascar and the Mascarene islands.

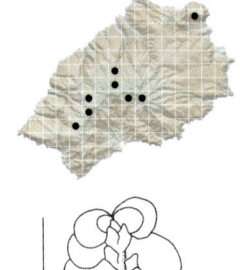

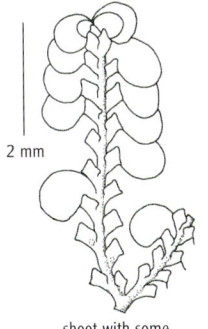

2 mm

shoot with some
deciduous leaves lost

LEFT **Moist shoots** TOP RIGHT **Perianths** BOTTOM RIGHT **Underside of shoot**

Frullania depressa

Frullaniaceae

Description Plants olive-green to blackish, prostrate, freely branched; leafy
shoots to 1.2 mm wide, 1–2 cm long. Leaves closely overlapping, lobes
0.6–0.8 mm long, the apices often reflexed. Leaf lobules helmet-shaped and
very variable, $^1/_4$–$^2/_3$ the size of the lobe, with the lower part partially
attached to the lobe. In shade, the lobule often develops into a flat, lanceolate
lamina, and a gradation in form can be seen in many specimens. Underleaves
rather large, shortly bilobed. Autoicous. Male inflorescences on very short
lateral branches. Female inflorescences at the end of main shoots or branches;
perianth with several keels and a prominent rostrum.

Recognition The genus *Frullania* is easily recognised by the small lobule that
forms a helmet- or lozenge-shaped 'water sac'. *F. depressa* is distinguished
(among other features) by the partial attachment of the lobule and the
perianths with several keels, and is the only member of the genus currently
known from St Helena. The character of the lobule is important for identifying
species, and forms from shaded habitats, with their atypical flattened lobules,
usually cannot be identified with certainty.

Habitats In mats on rocky outcrops and soil between boulders, either as 'pure'
colonies or mixed with other bryophytes, including *Cheilolejeunea ascensionis*
and *Trichostomum brachydontium*, and often with lichens; also on stones on
the woodland floor. Frequent in small quantities on the trunks and branches of native and non-native trees.

Status and distribution Widely distributed at middle to high altitudes, including some higher mist-exposed rock
outcrops in the drier areas along the south coast, but populations often small. Recorded from Diana's Peak
National Park, Distant Cottage, the Barn, the Boer Cemetery and Plantation House grounds. *Range*: Sub-
cosmopolitan in warm-temperate and tropical regions.

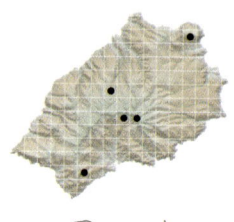

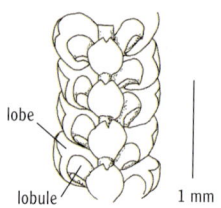

lobe

lobule

1 mm

underside of shoot

LEFT **Mat on tree trunk** RIGHT **Underside of shoot**

Aureolejeunea microscypha

Lejeuneaceae

Synonyms *Jungermannia microscypha, Lejeunea microscypha, Archilejeunea microscypha*

Description Plants deep brown to pale, dull brown (or greenish in shade), prostrate, forming small patches; leafy shoots 1.2–1.5 mm wide, compressed-cylindrical (rather worm-like), not strongly flattened against the substrate. Leaves 0.4–0.6 mm long, convex, the apex and margins of the lobe widely and strongly inrolled, and the lobule margin also strongly inrolled. Underleaves very broadly oval or kidney-shaped, 4–7 times the width of the stem, apex rounded to almost flat, not divided into lobes. Autoicous. Male inflorescences on short lateral branches; female inflorescences at the apex of a main stem or a lateral branch; perianth 1–1.5 mm long, bluntly keeled, with 1–2 branches from near its base.

Recognition The characteristic swollen, 'thick' appearance of the shoots, caused by the convex leaf lobes and lobules, and the strongly incurved apices and lower margins, contrasts with the more flattened shoots of other Lejeuneaceae on the island. The broad undivided underleaves are also highly distinctive. The only other species of the Lejeuneaceae on the island with undivided underleaves are *Aureolejeunea rotalis* and *Marchesinia brachiata* (see accounts for the differences).

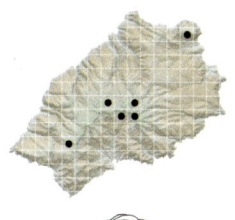

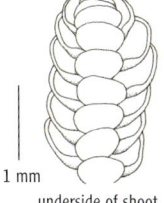

1 mm

underside of shoot

Habitats A mainly epiphytic species. In Diana's Peak National Park, on the living branches and decaying bark of whitewood, the smaller branches of black cabbage trees, and the trunk of he cabbage tree. Outside the national park, on the trunk of a large Monterey cypress (in abundance) and on a young Norfolk Island pine, and on the Haystack on damp, volcanic soil, between boulders with lichens and small kidney fern.

Status and distribution Endemic. Rather local and usually in small quantity. Found in the central part of Diana's Peak National Park; also recorded near Halley's Mount, Hooper's Rock, below Cason's, and the Barn.

LEFT **Tuft on tree branch** RIGHT **Underside of shoot**

Aureolejeunea rotalis

Lejeuneaceae

Synonyms *Jungermannia rotalis, Lejeunea rotalis, Phragmicoma rotalis*

Description Plants usually pale green, forming patches of flattened or somewhat convex shoots, or occurring in mixed bryophyte mats; stems pale; leafy shoots 1–1.6 mm wide and 1–2 cm long. Leaves closely overlapping, flattened to convex, lobes rounded, about as long as wide, 0.4–0.8 mm long, the margin flat or only slightly turned up; lobules narrowly rhomboidal, $^1/_3$ lobe length, inflated, the margin strongly inrolled. Underleaves rounded, 3–4 times the width of the stem. Autoicous. Male inflorescences on short lateral branches, usually several along the main shoot, often with up to 17 pairs of bracts; female inflorescences on lateral branches; perianth 1–1.5 mm long, bluntly keeled, with 1–2 leafy branches from near its base.

Recognition *A. rotalis* attracts attention when its generally pale shoots form extensive mats, as they often do. It is separated from all other related members of the Lejeuneaceae on the island, except *Aureolejeunea microscypha* and *Marchesinia brachiata*, by its undivided underleaves. However, those species are more robust: *A. microscypha* is further distinguished by its conspicuously inrolled lobe margins, broader underleaves and usually brownish colour; *M. brachiata* is a blackish- or olive-green plant, with flattened perianths.

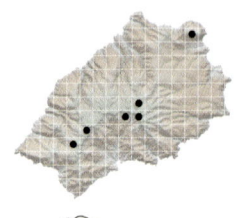

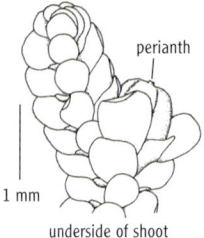

perianth

1 mm

underside of shoot

Habitats A mainly epiphytic species on the branches and sometimes trunks of native trees and tree fern, and occasionally non-native trees; frequent on the petioles of tree ferns, sometimes on the fronds of tongue-ferns, and other ferns. Also found growing in a *Campylopus* tuft in a rock crevice on the Barn, and on the cliffs of High Peak.

Status and distribution Endemic. Locally common in Diana's Peak National Park and other native upland habitats, often forming extensive mats on the branches of trees, especially black cabbage trees. One of the most frequent species of the Lejeuneaceae on the island.

perianths

LEFT **On tree branch** RIGHT **Fertile shoot**

Cheilolejeunea ascensionis

Lejeuneaceae

Synonyms *Jungermannia pterota, Jungermannia ascensionis*

Description Plants pale to olive-green, forming mats, or occurring as scattered shoots amongst other bryophytes. Shoots flattened, creeping, 1–1.5 mm wide and up to 3 cm long, occasionally branched. Leaves spreading from stem, slightly convex; lobes 0.5–0.7 mm long, the apex with a short, sharp point that is sometimes turned down in the moist leaf (occasionally the apex is merely bluntly pointed); lobules $^1/_3$ lobe length, inflated. Underleaves bilobed, rather large, 0.5 mm long. Autoicous. Fertile plants common; perianths with 5 short, blunt keels.

Recognition The pointed leaf apex distinguishes *C. ascensionis* from all other species of the Lejeuneaceae with underleaves found on the island, except for the much larger and dark coloured *Marchesinia brachiata*, in which the leaves are sometimes pointed. The leaf point is small and may sometimes be difficult to see in the field, especially when the leaf apex is turned downwards, but careful observation will reveal it, at least in some leaves.

Habitats A mainly epiphytic species of partial shade, growing on the trunks and branches of trees and shrubs, including native whitewood and black cabbage trees in Diana's Peak National Park, where it may be found in association with other species of Lejeuneaceae. It is also found on non-native trees elsewhere, including cypresses and Bermudan cedar, and is occasionally found on rock and damp soil between boulders, as at the Haystack, where it accompanies *Frullania* and lichens.

Status and distribution Endemic to St Helena and Ascension Island; fairly frequent, though often in small quantity, at mid to high altitudes.

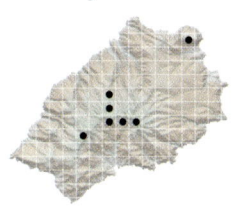

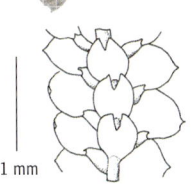

1 mm

underside of shoot

LEFT & TOP RIGHT **Fertile shots with perianths** BOTTOM RIGHT **Creeping shoots on petiole of tree fern**

Cololejeunea dianae **Lejeuneaceae**

Description A small, pale green liverwort forming compact mats or loose, straggling patches. Shoots sparingly branched, a few mm long and 1–1.3 mm wide. Leaves 0.5–0.7 mm long, when moist elevated from the substrate (occasionally becoming almost vertical in some shoots); lobules oval, inflated. Underleaves absent. Autoicous. Often abundantly fertile, male and female inflorescences borne on very short branches; perianths erect, with 5 prominent, spreading horns at the apex.

Recognition When sterile, this species may be recognised by the lack of underleaves, the moist leaves often angled upwards from the stem/substrate, and the broadly rounded lobes. When fertile, the perianths with 5 widely spreading horns (appearing star-like from above) immediately differentiates this species from all other small Lejeuneaceae on the island except *L. sanctae-helen*ae in which, however, the horns are sub-erect.

Habitats An epiphytic species, occurring in constantly moist and shaded conditions. It is most common on the decaying petioles of tree fern, where it may form extensive patches, or is found in mixed bryophyte communities. It can also occur on the living petioles of ferns and on the leaves of filmy fern. It appears to be uncommon on the smooth bark of native whitewood and non-native Cape yew and Mexican cypress trees.

Status and distribution Endemic. Widespread at upper altitudes in native habitats in Diana's Peak National Park, but rather rare and localised at the Depot, where it has been found only on non-native trees. It may occur elsewhere along the Central Ridge.

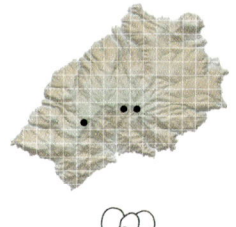

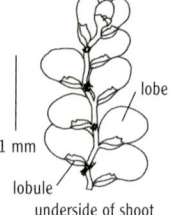

lobe

1 mm

lobule

underside of shoot

stylus

gemmae
on leaves

LEFT **Tuft and shoot on twig** TOP RIGHT **Underside of shoot** BOTTOM RIGHT **Gemmiferous leaves**

Cololejeunea grossestyla

Lejeuneaceae

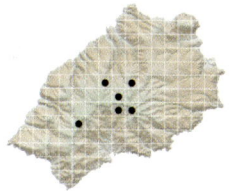

Description Plants small, pale to mid- or yellowish-green, appressed to the substrate, forming small patches or occurring as a few isolated shoots. Shoots flattened, up to 1.5 cm long and 1.5–2.8 mm wide, sparingly branched. Leaves 0.5–0.9 mm long, the lobes flat or weakly convex when dry and moist; lobules inflated, with an additional strap- or club-like appendage at the base, the stylus, at the base. Underleaves absent. Disc-like gemmae sometimes abundant on leaves. Autoicous. Male and female inflorescences borne on very short branches from the main stem; perianths with 5 long, sharp, narrow keels.

Recognition The stylus at the base of the lobule can just be seen in the field on close examination of the underside of the shoot with a ×20 lens, and this feature alone separates *C. grossesty*la from all other species on the island (and is a very rare feature in the genus worldwide). The sharply 5-keeled perianths will also distinguish this species from *C. dianae* and *C. sanctae-helenae*. Generally, lowland forms appear to be usually yellowish in colour, whilst forms at high altitudes are usually pale green.

Habitats A plant of shaded, sheltered locations, in a wide range of habitats, and seemingly tolerant of drying in some of its sites. At upper altitudes, found on the trunks and branches of native and non-native trees and shrubs; at mid-altitudes, it has been found on stones on the woodland floor, on the shaded, vertical face of a wall, at the base of an old stone gate-post, creeping over *Macromitrium urceolatum* on the trunk of a white olive, and on the twigs of shrubs.

Status and distribution Endemic. Rather widespread at middle and upper altitudes: recorded from Diana's Peak National Park, the Depot, the Sandy Bay Ridge road, Plantation House, Napoleon's Tomb. It is often in very small quantity and therefore easy to overlook.

LEFT **With** *Riccardia* sp. on bark RIGHT **Isolated shoot on bark**

Cololejeunea microscopica

Lejeuneaceae

Synonym *Aphanolejeunea microscopica*

Description Plants minute, pale to whitish-green or nearly colourless, forming thin, sparse mats or occurring as scattered shoots. Shoots up to 3–4 mm long and 0.25–0.5 mm wide, sparingly branched; stems translucent, thread-like. Leaves 0.25 mm long, usually well-spaced on the stem, spreading or ascending-erect from stem (and substrate), often appearing rather narrow in situ; lobes concave, ovate, tapering to a bluntly pointed apex; lobules about $^{3}/_{4}$ the length of the lobe, inflated, the apical cell forming a straight or slightly curved tooth. Underleaves absent. Autoicous. When fertile, the broadly rounded, proportionately large perianths are readily seen in the field (×20 hand lens).

Recognition This species can be recognised in the field by the minute size of the plant and the widely-spaced, concave, rather narrow, bluntly pointed leaves that are often raised up from the substrate. *Microlejeunea africana* and *Cololejeunea minutissima* are of similar size, but have more rounded leaves: all these species should be confirmed under the microscope.

Habitats A plant of well shaded, constantly humid environments. The most favoured habitat appears to be the decaying petioles of tree fern, either still attached or fallen to the ground, on which it often accompanies *Cololejeunea dianae*. It also occurs on the living petioles and laminae of filmy fern, and is sometimes found creeping over other epiphytic bryophytes on tree trunks.

Status and distribution On St Helena, fairly widespread in Diana's Peak National Park; also recorded at Hooper's Rock, and is presumably present elsewhere along the Central Ridge. *Range: C. microscopica* is widespread in western Europe, Africa, and in Central and South America.

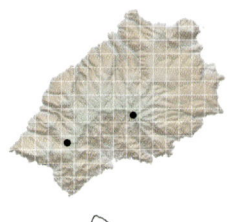

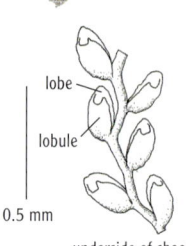

lobe

lobule

0.5 mm

underside of shoot

LEFT & TOP RIGHT **Tuft on vertical surface of stone wall** BOTTOM RIGHT **Fertile shoots**

Cololejeunea minutissima

Lejeuneaceae

Description Plants minute, pale to yellowish-green, forming tiny, straggling patches, or occurring as isolated shoots; leafy shoots 0.25–0.3 mm wide; stems very slender. Leaves 0.2 mm long, the lobes ovate or rounded, often distant on the stem and somewhat elevated from it; lobules more than half the size of the lobes; cells bulging-convex, margins strongly crenulate. Underleaves absent. Autoicous. Female inflorescences on short lateral branches; perianths often abundant, to 0.5 mm long, erect, with 5 blunt keels and thus appearing star-like from above.

Recognition This species is easily overlooked because of its minute size, especially in sub-optimal locations where it may be stunted. It can be mistaken for the similarly sized and rather more frequent *Microlejeunea africana* which, however, has underleaves (almost impossible to see in the field, and should be confirmed under the microscope). *Cololejeunea microscopica* is similar in size, but the plants are whitish-green or hyaline, and the leaves are ovate, not rounded.

Habitats In shaded locations: on the vertical face of a stone wall, on small stones on a woodland floor, and with *Fissidens translucens* on bare clay-loam soil on a woodland slope. Elsewhere in its range it is mostly an epiphyte, and on St Helena may perhaps occur also on the branches and trunks of trees, or creeping over other bryophytes.

Status and distribution On St Helena, apparently confined to middle altitudes, and thus far recorded from only a few places in the grounds of Plantation House and near the Boer Cemetery. *Range*: Sub-cosmopolitan, including the Cape Verde Islands and Macaronesia.

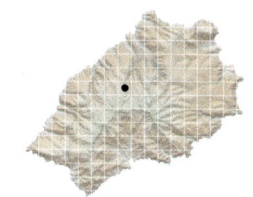

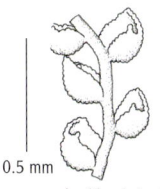

0.5 mm

underside of shoot

Shoots on branch of bilberry tree

Colura tenuicornis

Lejeuneaceae

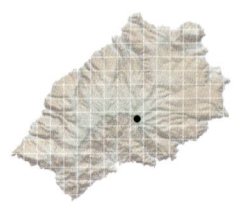

Description Plants resembling *C. calyptrifolia* in the pale green or yellowish colour, and in forming tiny patches or tufts on bark. Leaves 1–1.5 mm long, inflated into a rather narrow sac below and gradually contracted above to a long, horn-like beak which is $\frac{1}{3}$–$\frac{1}{2}$ the length of the leaf. Underleaves deeply 2-lobed. Autoicous. Perianths up to 1.5 mm long, with 5 spreading, long, horn-like keels at the apex (appearing star-like from above).

Recognition Species of *Colura* are easily recognised by the peculiar structure of the leaf, the beak being the most notable feature. *C. tenuicornis* differs from *C. calyptrifolia* in the rather narrower leaf sac that is gradually tapered into a longer narrow beak. Its form is usually quite distinctive in the field, but the length of the beak relative to the sac should be measured under the microscope to confirm its identity.

Habitats In small tufts on the twigs and small branches of whitewood, black cabbage tree, small fuchsia and bilberry tree, on the trunks of tree fern and on its petioles; also found on the stems of buck's-horn and on the fronds of filmy fern.

Status and distribution Known from only a few locations in Diana's Peak National Park. *Range*: Ascension Island, and widespread in Africa, South America, Asia and the Pacific Ocean region.

Fertile shoots with perianths on tree branch INSET Sterile shoot

Lejeunea autoica

Lejeuneaceae

Description Plants very small, mid- to pale green, forming small, loose patches or creeping over mixed bryophyte mats. Stems thread-like, with about 10 inner (medullary) cells (see Fig. 5, p. 9). Strong shoots 0.4–0.6 mm wide, but sometimes producing narrower, small-leaved branches. Leaves distant on the stem and spreading from it; lobes 0.25–0.35 mm long; lobules $^1/_3$–$^1/_2$ lobe length, strongly inflated, with a short, slightly curved tooth on the margin; the margin of the lobe forming an angle with the lobule keel. Underleaves bilobed to half way or more. Autoicous. Male and female inflorescences on short lateral branches; the male with 2–4 pairs of bracts, and the female producing relatively conspicuous perianths 0.4–0.5 mm long, with 5 keels.

Recognition This plant is recognised by its small size, leaves spreading from the stem, lobes much larger than lobules, and with an often distinct angle between the keel of the lobule and the margin of the lobe. It could be mistaken for the African *Microlejeunea kamerunensis* (not known from the island), especially when the shoots are weak and small. However, *M. kamerunensis* has a strongly curved tooth on the lobule, and the stem has only 3 inner cells (cross section).

Habitats On the base and branches of Mexican cypress trees at the Depot; on the decaying petiole of a tree fern, on the branches of Norfolk Island pine, and creeping over other bryophytes on the bark of a black cabbage tree in Diana's Peak National Park; creeping over *Frullania* and other bryophytes on the Haystack.

Status and distribution Recorded in Diana's Peak National Park, at the Depot and on the Haystack. *Range*: Known from west and east Africa, and Florida.

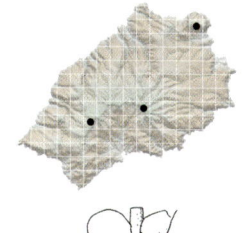

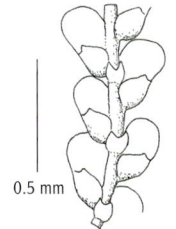

0.5 mm

underside of shoot

Creeping shoots on vertical rock outcrop RIGHT Fertile mat, with abundant perianths in central parts of the rosette

Lejeunea eckloniana

Lejeuneaceae

Description Plants medium-sized, pale bright green to dark green, glossy when dry, forming flat patches closely appressed to the substrate. Shoots 1.3–2 mm wide, sparingly branched. Leaves 0.7–1.1 mm long, lobes broadly and asymmetrically oval; lobules typically very small, composed of few cells, the apical cell forming a forward pointing tooth. Underleaves 2–3.5 times the width of the stem, deeply divided into two triangular lobes. Autoicous. Male and female inflorescences on very short branches. Perianths 0.5–0.7 mm long, with 5 long, sharp, equal keels; with a short branch at the base on which a second perianth may quickly develop (so perianths appear to be in close pairs).

Recognition The medium-sized plants, with flat shoots, rounded leaves and divided underleaves indicate a species of *Lejeunea*. The tiny leaf lobules (microscope) and, when fertile, the male and female inflorescences on very short branches on the same shoot are key features of *L. eckloniana*, and no other St Helena species of the Lejeuneaceae produces close 'pairs' of perianths. The glossy sheen of the dry plant can be a useful field character. Elsewhere in its range, plants of *L. eckloniana* may have larger leaf lobules and a stronger ochraceous or 'brassy' tinge.

Habitats On trees and rocks in usually deeply shaded, humid sites. On dry or moist, steep to vertical rock faces in woodland and by a waterfall, on small stones on the ground in a thicket of planted rebony, on wet boulders in a woodland stream, on tree roots and trunks, and on rotting wood.

Status and distribution Apparently local, mostly at middle altitudes: recorded from upper Deep Valley, Wash House Gut, the Depot, below High Peak, and by Francis Plain Gut. *Range*: Macaronesia, and widespread in tropical and southern Africa.

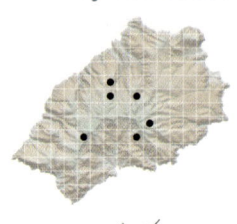

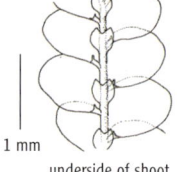

1 mm

underside of shoot

LEFT & CENTRE Upper side of shoots RIGHT Underside of shoot showing inflated lobules

Lejeunea sanctae-helenae

Lejeuneaceae

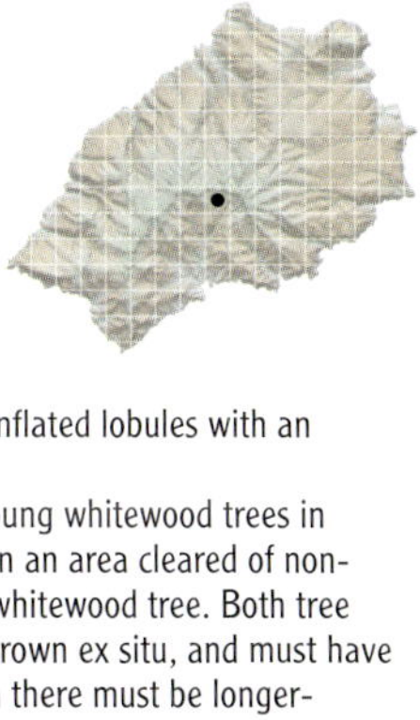

Description Plants small, pale to mid-green or somewhat yellowish-green; shoots appressed to the substrate, 2–2.3 mm wide, freely branched. Leaves 1.1–1.5 mm long, the lobes broad and apex broadly rounded; lobules rounded, strongly inflated, hood-like, with the margin strongly inrolled. Underleaves about the same size as the lobules, bilobed. Autoicous, the male and female inflorescences borne on short branches; perianths to 0.9 mm long, with 5 short keels.

Recognition The flattened shoots with appressed, rounded leaves are features of most of the other species of the Lejeuneaceae on the island. However, *L. sanctae-helenae* is easily recognised in the field by the extraordinarily strongly inflated lobules with an inrolled margin that are unlike those in any other St Helena species.

Habitats Found only on the smooth bark of the branches and trunks of two living, young whitewood trees in moderate shade, and on the trunk of a young he cabbage in fairly open locations in an area cleared of non-native vegetation (mostly flax); also on the rotting bark of a young but moribund whitewood tree. Both tree species are endemic to St Helena. However, since all the trees had been nursery-grown ex situ, and must have been colonised by *L. sanctae-helenae* after they were planted out, it seems certain there must be longer-established, undiscovered populations of the liverwort elsewhere on the island producing the spores for the colonisation of the young trees.

Status and distribution Endemic. Apparently very rare and in small quantity, and currently known only within a 50 × 50 m area of Diana's Peak National Park.

INSET **Tuft on bark of Mexican cypress** RIGHT **Fertile shoots with perianths**

Marchesinia brachiata

Lejeuneaceae

Synonyms *Jungermannia acutiloba, Phragmicoma acutiloba*

Description Plants strongly pigmented, the youngest part of the shoots mid-green; older parts dark olive- or blackish-green, glossy, frequently branched; leafy shoots creeping, flattened against the substrate, 2–3 mm wide, forming discrete, spreading colonies on bark. Leaves overlapping, slightly convex, 1–1.8 mm long; lobes broadly ovate, the apex rounded or shortly pointed; lobules $^1/_3$ the length of the lobe, with 2–4 small, prominent marginal teeth. Underleaves large, rounded, not divided into lobes, $^1/_4$–$^1/_2$ width of shoot. Dioicous or autoicous. Perianth at the apex of a main shoot or a lateral branch, but with 1–2 branches arising from near its base and eventually extending beyond it. Perianths conspicuously flattened, 2.5 mm long, with 2 sharp lateral keels, apex flat or slightly indented, with a short rostrum.

Recognition The predominantly dark colour and large size differentiates this species at a glance from all other species of the Lejeuneaceae on the island. When fertile, the flattened perianths with the flat apex are also highly distinct. The prominent lobule teeth can be seen on close examination with a ×20 hand lens.

Habitats Currently known only from the trunks and branches of Mexican cypress trees and on soil among rocks on a grassy slope. In 1839, Hooker reported this species from branches of black cabbage trees on Diana's Peak.

Status and distribution This species is locally frequent in a small area of woodland at the Depot and nearby but, despite recent searches, has not been found again on black cabbage trees. *Range*: *M. brachiata* is otherwise known only from Central and South America, where it is widespread.

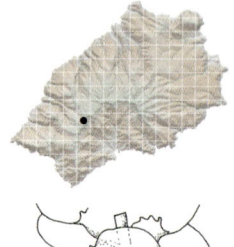

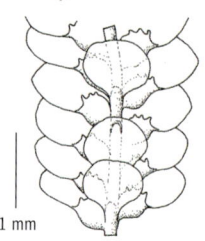

1 mm

underside of shoot
with one underleaf removed
to show lobule

TOP LEFT **Dry shoots creeping over** *Frullania depressa* BOTTOM LEFT & RIGHT **Moist shoots**

Microlejeunea africana

Lejeuneaceae

Description Plants very small, pale or whitish-green when dry, more translucent when moist, forming small, diffuse patches or frequently occurring as individual shoots creeping over and amongst other bryophytes. Leafy shoots to 0.25–0.35 mm wide, stems thread-like. Leaves distant on the stem, generally lying parallel to it and sometimes elevated from the substrate; lobes 0.15–0.2 mm long; lobules proportionately large, inflated, at least $^1/_2$ the size of the lobe, with a long, curved tooth on the margin; the margin of the lobe forming an almost straight line with the lobule keel. Underleaves a little wider than the stem, bilobed (microscope). Dioicous. Perianths 0.2 mm long, broadly ovate, with 5 keels.

Recognition This species is recognised by its very small size, leaves distant on the stem and parallel to it, and by its divided underleaves. *Cololejeunea minutissima* and *C. microscopica* are similar in size to *M. africana*, but lack underleaves (microscope). *Lejeunea autoica* is also small and pale in colour, but is larger, the leaves spreading widely from the stem and the leaf lobes much larger than the lobules.

Habitats A mostly epiphytic species, but also found closely appressed to hard stones in Plantation Wood. It has been found on the trunks and branches of whitewood, he cabbage and bilberry tree in Diana's Peak National Park, and on white olive and acacia/wattle species elsewhere. It grows directly on the bark, and epiphytically over and through mats of larger bryophytes, sometimes only as isolated shoots.

Status and distribution Frequent in Diana's Peak National Park, and present elsewhere along the Central Ridge and at lower altitudes. Easy to overlook, and certain to be found in many other locations. *Range*: Widespread in tropical and southern Africa, including the Indian Ocean islands.

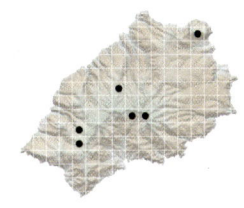

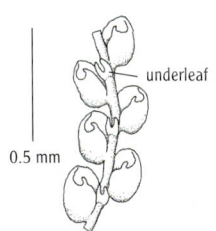

underleaf

0.5 mm

underside of shoot

LEFT **Typical weft from shady habitats** RIGHT **Upper side of shoots**

Bazzania praerupta

Bazzaniaceae

Description Plants mid-green, the older parts often olive-green or somewhat brownish, forming loose mats; leafy shoots with a rather soft texture, 1.4–2 mm wide, stems green. Branching frequent, regularly forked or unequal; in addition, with frequent flagelliform branches arising at a wide angle from the underside of the stem. Leaves 0.7–1.1 mm long, widely spreading from stem, somewhat curved, usually distant on the stem, sometimes deciduous, apex very variable, typically shallowly divided into 2–3 sharp teeth, but sometimes teeth obscure, or absent. Underleaves distant, rounded, sometimes weakly toothed at the apex. Fertile plants not found.

Recognition Easily recognised by the 2 to 3-toothed apex on many leaves, the forked branching, the rather rigid flagelliform branches arising at a wide angle from the underside of the shoot, and its general habit.

Habitats A plant of usually rather deep shade, on banks and steep slopes under a canopy of tree ferns or cabbage trees, and presumably requiring high humidity. On peat/humus and decaying vegetation, on the rotting bases of tree ferns, and in one place, on the horizontal trunk of a black cabbage tree.

Status and distribution Known only from Diana's Peak National Park, where it appears to be rather scarce, though it forms fairly extensive mats in one or two places. The St Helena plants are slender and atypical of the species, and are presumed to be small, shade forms. *Range*: *B. praerupta* is known from tropical Africa, Madagascar, the Mascarene islands, the Indian subcontinent and south-east Asia.

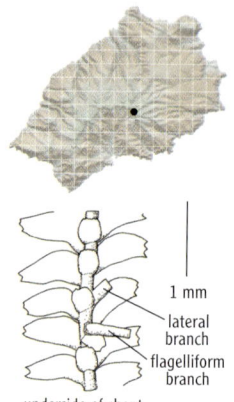

1 mm
lateral branch
flagelliform branch
underside of shoot

TOP LEFT **Typical form with appressed leaves** BOTTOM LEFT **Shade form with more spreading leaves** RIGHT **Shoots showing leaf form**

Kurzia nemoides   Lepidoziaceae

Synonym *Jungermannia nemoides*

Description Plants olive- or brownish-green (pale in very deep shade), forming mats or hummocks or intermixed with other bryophytes. Leafy shoots 0.2–0.3 mm wide, stems thin, wiry. Leaves 0.2–0.25 mm long, deeply divided into 4 (sometimes 3) narrowly triangular lobes, the leaf base widely spreading from the stem, but the lobes sharply angled upwards, thus becoming sub-erect. Underleaves deeply divided into 3–4 segments. Dioicous. Perianths apparently rare, proportionately very large, 1.5–2 mm long, cylindrical, tapered above, the mouth fringed with hair-like teeth. Male plants not yet found.

Recognition The brownish mats or hummocks, and slender, wiry shoots with tiny, deeply segmented leaves easily distinguish this species from all others on the island. Although large and easy to see with a hand lens, the rarity of perianths makes them difficult to find in the field. The only other species with leaves deeply divided into segments is *Lepidozia africana*, but that species differs in many features including habit, colour, and leaf characters.

Habitats A plant of moist, shaded environments, and found mainly in the native cloud forest communities: on living and fallen trunks of tree fern and black cabbage tree, on decaying tree fern debris, humus-rich banks and slopes under vegetation canopy, and soil overlying rocks. In the more open habitats, it forms compact mats or hummocks; in deep shade, colonies can be more diffuse with more loosely interweaving stems.

Status and distribution Endemic. Widespread and frequent at high altitudes throughout Diana's Peak National Park, with a single outlying record from the Depot. It may also occur at High Peak, and perhaps elsewhere along the Central Ridge.

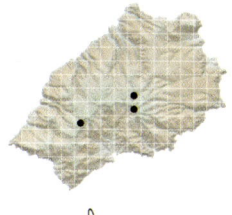

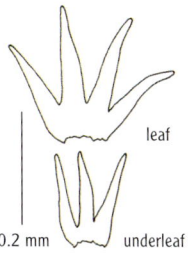

leaf

0.2 mm   underleaf

Shoots with male inflorescences

male inflorescences

Chiloscyphus humistratus

Lophocoleaceae

Synonyms *Jungermannia humistrata, Lophocolea humistrata*

Description Plants usually pale green, but occasionally darker green in open locations, forming low mats. Leafy shoots 1.5–2.5 mm wide, flattened, except at the shoot apex where opposite pairs of the youngest leaves are often folded together and elevated from the substrate. Leaves to 1 mm long, slightly overlapping, the apex rounded, truncate or shallowly indented (occasionally some weakly 2-lobed leaves present), margins sometimes crenulate. Underleaves small, deeply 2-lobed. Autoicous. Male and female inflorescences at apex of main shoots or on short lateral branches; perianths to 1.8 mm long, with 3 sharp, toothed keels.

Recognition This distinctive species is recognised by the usually pale hue, flattened shoots, sub-rectangular leaves and 2-lobed underleaves. The other named species of *Chiloscyphus* on the island, *C. coadunatus*, is easily differentiated by its leaves divided into two long, slender lobes.

Habitats A plant mostly of moist, shaded habitats, but sometimes growing in exposed locations. Most records are from native habitats at high altitudes in Diana's Peak National Park, where it is found on the trunks and bases of trees and tree ferns, on rocks and stones, on rock scars, peaty soil banks, and on decaying vegetation. Outside the national park, it has been found on the moist, spongy bark of a mature *Eucalyptus*, and on soil on steep-cut roadside banks.

Status and distribution Endemic. Mainly found in Diana's Peak National Park, but there is a scatter of recorded sites elsewhere, including the Depot, and by the Sandy Bay Ridge road.

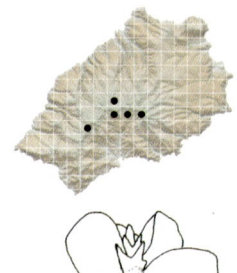

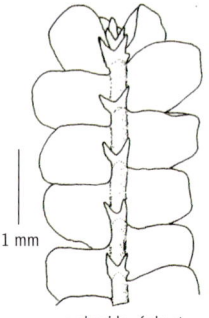

1 mm

underside of shoot

TOP LEFT Creeping shoots BOTTOM LEFT Shoot showing toothed leaves RIGHT Deciduous shoot with some leaves already detached

Plagiochila spinulosa

Plagiochilaceae

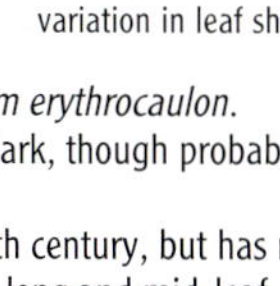

Description Plants dark green, procumbent, in loose, mixed bryophyte mats. Leafy shoots 1.4–1.7 mm wide, sparingly branched, small-leaved branches often present; leaves often deciduous, and stems and branches often partly denuded. Leaves usually rather distant, 0.9–1 mm long when well developed, the front margin lacking teeth, reflexed and rather long-decurrent down the stem, the back margin convex, usually with 3–8 spine-like teeth, the apex rounded or truncate, with 2–3 long teeth. Leaf cells rounded, 20–30 μm long, with large or small trigones. Cuticle ornamented with short, fine lines or papillae. Fertile shoots not observed.

Recognition The small size, the leaves with spinose teeth on the back margin and apex, and the partly denuded stems and branches are good field characters. However, mats may include many weak shoots bearing small (0.6–0.8 mm), anomalous leaves that may be divided into 2 acute lobes, with few or no spinose marginal teeth. This species (including the weak, anomalous form) is provisionally named *P. spinulosa* pending further study.

Habitats In usually deeply shaded, constantly humid sites under a canopy of native trees and ferns, on the trunks of trees, rotting vegetation, and in hanging 'sheets' of bryophytes on steep banks. In mixed mats, common associates include *Chiloscyphus humistratus*, *Fissidens chioneurus* and *Sematophyllum erythrocaulon*.

Status and distribution Found in a single 1km grid square in Diana's Peak National Park, though probably present elsewhere. *Range*: Western Scandinavia, Atlantic Europe and Madeira.

Further species Another species, *P. insularia*, was collected by J.J. Haughton in the 19th century, but has not been recorded since. It is much larger, the shoots to 4.5 mm wide, leaves to 2.5 mm long and mid-leaf cells 40–65 μm long. The habitat is not known. Although *P. insularia* has not been reported from elsewhere, further collections on St Helena may show it to be synonymous with other known species from Africa or South America.

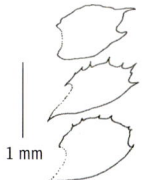

1 mm

variation in leaf shape

Sterile shoots in a mixed bryophyte mat

Adelanthus decipiens

Adelanthaceae

Synonyms *Jungermannia campylodonta, Plagiochila campylodonta*

Description Plants green to dark green, becoming dark brown or blackish on drying, forming mats or cushions. Leafy shoots erect, 5–15 mm long, arising from much-branched, creeping stems; microphyllous branches occasional to frequent; young shoots often rather flattened, with leaves nearly erect and folded together. Leaves rounded or oval, 0.6–1.1 mm long, obliquely set on the stem and partially clasping it (and therefore distinctly concave), spreading to erect, apex variable, typically very shortly bilobed or with 2 distant, coarse teeth that may be sharp or blunt. Front margin of leaf often narrowly incurved (at least near base) and decurrent at base. Underleaves absent. Fertile plants not seen.

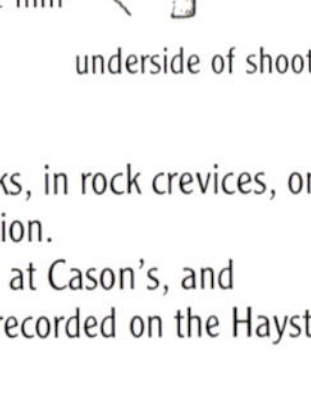

Recognition This species can be recognised by the concave, obliquely-set leaves, the leaf apex bilobed or with 2 small teeth, and the narrowly turned up front leaf margin. Small or weak shoots may not show the typical characters of the species, so well developed shoots in a tuft should be examined. Of the other species with which some growth forms of *A. decipiens* might be confused, *Plagiochila spinulosa* has leaves that are not concave and usually spinose-toothed along the back margin, and *Tylimanthus anisodontus* differs in its leaf shape and other characters (see account, p. 51).

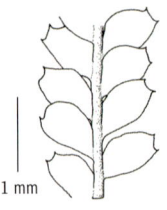

1 mm

underside of shoot

Habitats In mostly shaded, moist sites: on humus-rich and mineral soils, soil-capped rocks, in rock crevices, on moist bark, and roots of living trees and tree ferns, on rotting trunks and other vegetation.

Status and distribution Frequent in Diana's Peak National Park, local at the Depot and at Cason's, and presumably present elsewhere in suitable habitats along the Central Ridge; otherwise recorded on the Haystack. *Range*: Europe (local), Macaronesia, Africa, and Central and South America.

LEFT Creeping over *Phaeoceros carolinianus* RIGHT Herbarium specimen having lost its green pigment

Cephalozia sanctae-helenae

Cephaloziaceae

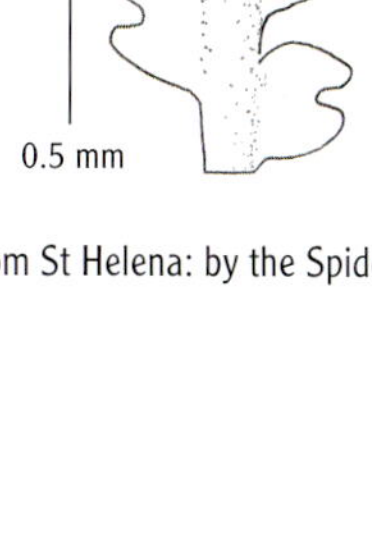

Description Plants mid or pale green, rather fleshy in appearance, forming diffuse patches or occurring as single shoots creeping amongst other bryophytes; leafy shoots 0.5–0.8 mm wide, stems broad (to $^1/_3$ width of shoot), cortical cells 45–70 µm wide. Leaves obliquely inserted, 0.2–0.3 mm long, decurrent at base, $^1/_4$–$^1/_3$ divided into two short, bluntly rounded lobes that sometimes slightly converge. Leaf lamina cells large, 30–40 µm wide, thin-walled; walls of apical cell of lobe evenly thickened (not thickened at the apex). Underleaves absent. Autoicous. Perianths on very short lateral branches, 1.5 mm long, the mouth crenulate; female bracts lobed, not toothed.

Recognition This small, but distinctive species can be recognised in the field by the nearly longitudinally inserted, bilobed leaves with blunt lobes, large cells (hand lens) and the rather fleshy appearance. Leaves of *Cylindrocolea helenae* are more rectangular, with longer, more acute lobes and smaller cells. *Cephaloziella* spp. are smaller, with transversely inserted leaves, pointed leaf apices and small cells.

Habitats In Diana's Peak National Park, near the base of a stony earth bank, creeping over *Phaeoceros carolinianus*, and associated with *Dicranella proscripta*, *Riccardia* sp., *Cephaloziella* sp. and *Chiloscyphus coadunatus*. Also on the south-facing cliff of High Peak, associated with *Lepidopilidium pallidifolium*. In both places, it was in small quantity.

Status and distribution Apparently endemic. Currently, there are only two records from St Helena: by the Spider Sprint path in Diana's Peak National Park and on High Peak.

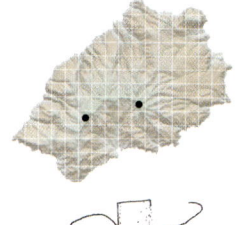

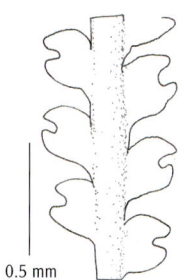

0.5 mm

Sterile shoots of two different species

Cephaloziella spp.

Cephaloziellaceae

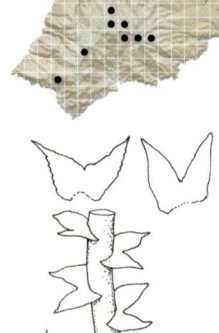

0.2 mm

Description Very small or minute plants, green or brownish, sometimes tinged purple, forming patches or in mixed bryophyte mats; stems thread-like, green to blackish. Shoots 0.2–0.5 mm wide. Leaves distant, spreading or reflexed, transversely set on the stem, 0.1–0.2 mm long, deeply divided into two acute lobes, the lobes toothed or not. Underleaves, if present, minute (not visible with a hand lens). Male and female inflorescences at the apex of main shoots or branches; bracts larger than the leaves, often toothed; perianths usually cylindrical, up to 1.5 mm long, and relatively conspicuous. Gemmae are sometimes produced at the shoot apex.

Recognition The thread-like stems, and minute, bilobed, transverse leaves are key characters. Although perhaps 3–4 species have been collected from St Helena, the genus has been little studied in tropical and sub-tropical areas, and it has not been possible to name them at the present time. With experience, it is sometimes possible to recognise the different forms in the field with a ×20 lens, but microscopic examination of fertile material is almost always essential for species identification, so this should be collected at every opportunity. However, even with fertile material, determination is technically difficult.

Habitats Mainly plants of freely draining, gritty or fine-grained mineral soils in open habitats or shaded by vegetation: on earthy banks by roads and paths, in earthy crevices at the base of rock outcrops, on thin soil overlying rocks.

Status and distribution Species of *Cephaloziella* are fairly widespread at upper and middle altitudes on the island, but are apparently absent from the cloud forest communities. *Range*: The genus is cosmopolitan.

LEFT **Robust form, with perianths** RIGHT **Weaker form**

Cylindrocolea helenae

Cephaloziellaceae

Description Plants minute or small, pale green to dark green, forming diffuse patches flattened to the substrate, branched, variable in size and habit. Shoots 0.3–0.6 mm wide, but weaker in deep shade; stems delicate. Leaves 0.18–0.35 mm long, longitudinally inserted on the stem, decurrent, usually widest above the base, divided to half way into two acute lobes. Underleaves absent. Autoicous. Male and female inflorescences at the end of leading shoots, or on short branches. Male spikes with 5–9 pairs of bracts. Female bracts very widely spreading, and reflexed; perianths almost cylindrical, 0.7 mm (in deep shade) to 1.2 mm long, the mouth crenulate or rather ragged.

Recognition Although this species is variable (plants in deep shade being pale green and small, whilst better illuminated plants are deeper green and more robust), it can be recognised in the field by its tiny size and the flattened shoots with longitudinally inserted, decurrent, bilobed leaves. Perianths are relatively conspicuous, and sub-erect from the substrate. Microscopically, the often rather ragged appearance of the perianth mouth is distinctive. Small, weak forms are very inconspicuous in the field. Species of *Cephaloziella* have transversely inserted leaves.

Habitats In deeply to moderately shaded, humid places on a wide range of substrates: on trunks of living tree ferns and other trees (sometimes creeping amongst and over *Sematophyllum* or other bryophyte species), on petioles and laminae of tree ferns, on rotting wood, on soil overlying rock and in soil-filled crevices.

Status and distribution Presumed endemic. Recorded in Diana's Peak National Park (weak forms from deep shade), the Depot, Blue Hill Village, the Boer Cemetery and Hutt's Gate.

0.5 mm

weak and robust shoots

LEFT **Isolated shoots growing through** *Kurzia nemoides* TOP RIGHT **Upright tuft** BOTTOM RIGHT **Gemmiferous shoots**

Anastrophyllum subcomplicatum

Scapaniaceae

Synonym *Jungermannia obtusata*

Description Plants small, variable, pale to reddish-brown, magenta or even dark purple-black, forming small, compact, pure or mixed tufts or mats, shoots 0.6–1 mm wide, to 2 cm long. Leaves 0.9–1.2 mm long, strongly concave, asymmetrically rounded, widely spreading from stem and somewhat curved to one side, part of the leaf base transversely inserted on the stem, and part obliquely inserted; intact leaves on non-gemmiferous shoots typically deeply bilobed, the acute lobes ending in spinose points; leaf lobes eroded on gemmiferous shoots. Leaf margin sometimes with a sharp tooth near the base. Sometimes many or most shoots in a tuft producing knobbly gemmae from the apex of the youngest leaves, forming a gemma cluster at the shoot apex. Underleaves absent. Fertile plants not seen.

Recognition This small species has a distinctive habit, with the deeply concave leaves clasping the stem. Leaves that are obviously deeply bilobed with sharply pointed leaf apices may be difficult to find, since many shoots may be gemmiferous, and the leaf apices eroded and rounded, or lacking. The leaves are easily broken or split when detaching them for microscopic examination.

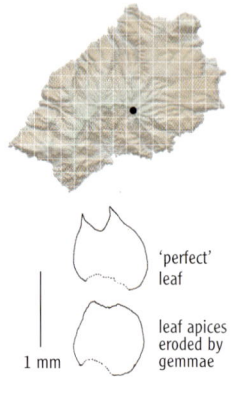

'perfect' leaf

leaf apices eroded by gemmae

1 mm

Habitats Found on the trunks of tree ferns and black cabbage trees in moist, shaded locations: in tufts or in mixed bryophyte communities including *Kurzia nemoides, Sematophyllum erythrocaulon, Aureolejeunea rotalis* and *Chiloscyphus humistratus*. In 1844, Hooker reported it "amongst tufts of moss at the roots of trees, Diana's Peak".

Status and distribution Very localised and in small quantity in Diana's Peak National Park. *Range:* Otherwise reported only from Tristan da Cunha, and from the Mariana Islands and Caroline Islands in the north Pacific. This extraordinary disjunction in range has led some taxonomists to consider that two different species are involved.

Mat of sterile shoots

Tylimanthus anisodontus

Acrobolbaceae

Synonyms *Jungermannia anisodonta, Plagiochila anisodonta, Tylimanthus anisodon*

Description A small, rather delicate liverwort, bright green to dull olive green, becoming darker when dry, forming loose masses or mats; leafy shoots 2–5 mm wide, sparsely branched. Leaves distant to somewhat overlapping on the stem, 1–2.5 mm long, widely spreading, somewhat down-curved, divided at the apex into 2 unequal, acute lobes. Cells irregularly hexagonal, the leaf cuticle rough with elongate raised areas (verrucae) or short, minute ridges. Underleaves very minute and inconspicuous, only 1–3 cells long. Fertile plants not seen.

Recognition *T. anisodontus* is differentiated from the other liverworts on St Helena that have bilobed leaves by the very unequal lobes and the verrucose cuticle that gives the leaf surface a rough appearance under a microscope at high magnification (×400). However, the verrucae may be difficult to see on some leaves. Weak forms of *Adelanthus decipiens* with distant leaves can be somewhat similar in general form, but the leaves are rounder, toothed rather than divided, at least weakly concave, long-decurrent and have a smooth cuticle.

Habitats Grows in small, thin, diffuse mats on a moist, very shaded vertical rock face under a deep overhang by a seepage line. Tall vegetation, including tree fern grows adjacent to the rock face, thus maintaining a shaded, constantly humid environment.

Status and distribution Endemic, confined to Diana's Peak National Park, and apparently very rare. It is currently known from only one site, where it grows patchily over an area of only about 40 × 30 cm. It was reported in the 19th century, but the location is unknown.

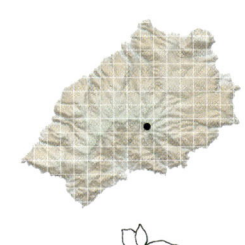

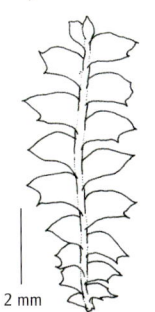

2 mm

LEFT **Green form** RIGHT **Brownish form**

Mnioloma fuscum

Calypogeiaceae

Synonym *Calypogeia fusca*

Description Plants growing in prostrate mats, usually brownish with the youngest parts of the shoot green, but sometimes wholly pale to dark green; stems usually brown. Leafy shoots 1–2 mm wide, sparingly branched, sometimes with microphyllous branches. Leaves broadly oval to oblong-oval, 0.5–1 mm long, apex rounded, truncate or slightly notched, the margins crenulate (often the marginal cells conically bulging, with thickened points). Underleaves rounded, rather small, the margins toothed or crenulate, the apex rounded or slightly notched. Fertile plants not observed on St Helena.

Recognition On cursory inspection in the field, green forms might be mistaken for a species of the Lejeuneaceae. However, *M. fuscum* is readily separated by the lack of a lobule on the leaf, the crenulate leaf margin (which can just be seen in some shoots with a ×20 hand lens), the rather small, rounded underleaves and usually by its brownish colour. Once known, it is an easy plant to recognise in the field.

Habitats On moist rock and humus-rich soils in deeply shaded sites under a canopy of tree fern or black scale fern. Associates include *Fissidens chioneurus*, *Lepidopilidium pallidifolium*, *Philonotis helenica*, *Riccardia* sp. and *Syrrhopodon gaudichaudii*.

Status and distribution Known only from high altitudes in the central part of Diana's Peak National Park where, however, it is fairly frequent in small patches in the native tree fern thicket. *Range*: The Azores, Tristan da Cunha, tropical and southern Africa, eastern Asia and the Pacific Ocean region.

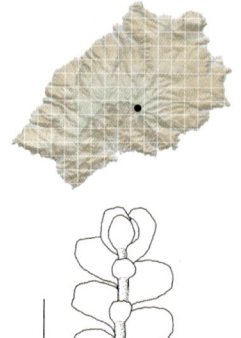

1 mm   underleaf
underside of shoot

LEFT **Convex and flat forms of the thallus** RIGHT **Sporophytes**

Exormotheca pustulosa

Exormothecaceae

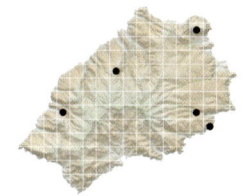

Description Thallus glaucous-green to silvery-white, small, once or twice-forked, the branches 2 mm wide; upper surface of thallus very rough with numerous steeply conical swellings, each with a pore at the apex. Large scales on underside of thallus whitish with some purple coloration, in 2 rows, with or without slender appendages. Thallus with air chambers in one layer (cross-section), with numerous green filaments inside. Monoicous. Male receptacles embedded in a furrow along the median thallus-line. Female receptacles 1 to 2-lobed, elevated at maturity on a pale stalk 4–16 mm high, producing either a single capsule, or 2 capsules on opposite sides of the receptacle.

Recognition *E. pustulosa* is very easily recognised by the small, whitish thallus with conical swellings, this alone being sufficient to distinguish it from all other species on the island. Fertile plants are also highly distinctive, the stalked receptacles with 1 or 2 round, black capsules also clearly differentiating this species from all other thallose species on St Helena.

Habitats A plant of dry environments, forming crowded patches on freely draining soils in open places, on soil capping on rock outcrops, in sheltered dry crevices of rocks and on paths and dry banks among shrubs.

Status and distribution This species mostly occurs at mid altitudes on the island in seasonally parched communities. It is locally abundant with *Bryum argenteum* and *Trichostomum brachydontium* on paths and among shrubs on the west-facing slopes of High Knoll Fort, and common on rocky hill-slopes across much of the island, including Great Stone Top, the Haystack and Spyglass Ridge. *Range*: Widespread, but local, in warm-temperate Africa and Europe, including the Cape Verde Islands and Macaronesia.

Different growth forms BOTTOM RIGHT **Thallus showing marginal teeth**

Jensenia spinosa

Pallaviciniaceae

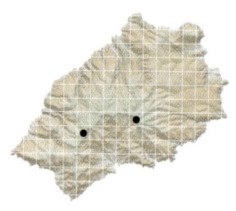

Synonym *Symphyogyna serrata*

Description Plants medium-sized, mid-green, erect, in crowded tufts; lower part of thallus narrow, stalk-like, arising from a creeping rhizome; thallus forked once or twice into flat thallus branches 1–2 mm wide, each with a broad midrib, the thallus wings slightly undulate, strongly toothed. Dioicous. Male reproductive structures globose, covered by scales, scattered along the upper surface of the midrib. Archegonia in small groups at or just behind the bifurcation (division) of the frond, protected by a small cup-like outgrowth (involucral scale). After fertilisation, a broadly cylindrical calyptra-like structure a few mm long, fringed with hair-like segments around the top, encloses the developing capsule, from which the seta and capsule emerge. However, neither male nor female structures have yet been observed on St Helena plants.

Recognition *J. spinosa* is easily recognised by its upright habitat, the branched thallus raised on a stalk and the coarsely toothed thallus branch margins (easily observable in the field with a hand lens). No other thallose liverwort on St Helena shows these features. The thallus margins of *Symphyogyna brasiliensis* are not toothed. Fertile plants are distinguished by the cup-like involucre that protects the female structures; the involucre is a flap or scale in *Symphyogyna*.

Habitats Damp rock faces and crevices, and moist soil banks, in moderately to deeply shaded, humid places under tree ferns and other vegetation.

Status and distribution Known from several locations in Diana's Peak National Park, and from High Peak. *Range*: Widespread in tropical and sub-tropical Africa and South America.

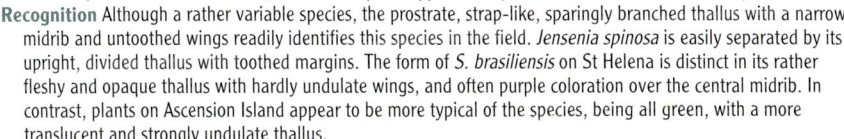

Branching population growing through *Kurzia nemoides* INSET Thallus with involucral scales

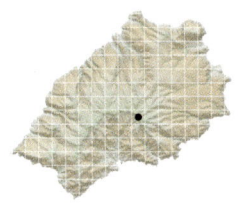

Symphyogyna brasiliensis   **Pallaviciniaceae**

Synonyms *Jungermannia procumbens, Pallavicinius procumbens*

Description Thallus strap-like, arising from rhizomes, prostrate, green, simple, but frequently with branches arising from the nerve on the underside of the thallus 3–6 mm wide, margins lacking teeth, weakly undulate or almost flat; midrib narrow, often conspicuous with purplish coloration, up to 10 tiers of cells thick in the centre, grading rather abruptly to wings 1 cell thick. Dioicous. Male reproductive structures crowded over the midrib, each covered by a scale. Archegonia in small groups over the midrib, each protected by a small scale; after fertilisation, a sub-cylindrical or club-shaped, fleshy calyptra about 5 mm long develops and encloses the developing capsule, and from which the seta and capsule emerge. However, neither male structures, nor a developed calyptra have yet been observed on St Helena plants.

Recognition Although a rather variable species, the prostrate, strap-like, sparingly branched thallus with a narrow midrib and untoothed wings readily identifies this species in the field. *Jensenia spinosa* is easily separated by its upright, divided thallus with toothed margins. The form of *S. brasiliensis* on St Helena is distinct in its rather fleshy and opaque thallus with hardly undulate wings, and often purple coloration over the central midrib. In contrast, plants on Ascension Island appear to be more typical of the species, being all green, with a more translucent and strongly undulate thallus.

Habitats In moderately to deeply shaded places, on peaty or humus-rich soil on banks, vertical or steep rock faces or in crevices, occasionally creeping over a moss carpet or growing on decaying vegetation. Also found growing on the base of a tussock of Diana's Peak grass.

Status and distribution Known only from Diana's Peak National Park and from High Peak. *Range*: Ascension Island, and widespread in tropical and sub-tropical Africa, and South America.

Three different species (yet to be named)

Riccardia spp.

Aneuraceae

Description Plants thallose, dark or deep green to mid- or yellowish-green, freely 1 to 2-pinnately branched, the branches strap-shaped or linear, at a wide angle (70–90°) from the main axis. Thallus branches not more than 1.5 mm wide, narrowly ellipsoid in cross-section (3–6 cells thick in the middle); in some species, the branches have a translucent margin one cell thick and several cell rows wide. Cells large and thin-walled. Male and female structures on short, specialised, lateral branches. Male branches with up to 15 pairs of antheridia in two rows. Calyptra 2–5 mm long, club-shaped or cylindrical, fleshy; capsule cylindrical or ellipsoidal, on a long, short-lived seta.

Recognition The genus is easy to recognise in the field by the narrow, much-branched thallus, but the species can be technically difficult to identify, and microscopic examination is invariably required. Furthermore, the morphology can vary in response to different environmental conditions, though this is poorly documented. A preliminary investigation shows there may be 3–4 species of *Riccardia* on St Helena, but at the present time, it is not possible to name them because the genus is so poorly known in many areas, including in much of Africa.

Habitats In shaded, constantly moist or wet habitats. On humus-rich mineral or peaty soils, steep rocky soil banks, wet rocks by streams and the damp base of a roadside wall. Also locally frequent on moist trunks and branches of living and dead tree ferns and other native and alien trees, and on rotting vegetation.

Status and distribution Apparently confined to the Central Ridge, where precipitation maintains the required moist conditions. *Range*: The genus is cosmopolitan and common in suitable habitats.

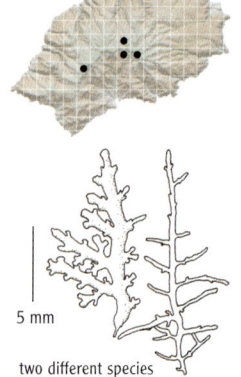

5 mm

two different species

LEFT Plants with dehiscing capsules RIGHT Plants with undehisced capsules, showing the cup-shaped, crisped thalli

Anthoceros sp.

Anthocerotaceae

Description Thallus bright green, to 5 mm wide, shortly lobed, the margins ascending, undulate-crisped; thallus with large internal cavities filled with mucilage. Each thallus cell with a single, very large chloroplast. Monoicous. Antheridia in cavities in the upper surface of the thallus. Capsule green, linear, stick-like, to 4 cm long, blackening from the apex at maturity and dehiscing (splitting) along two suture lines to release black spores.

Recognition The only species with which *Anthoceros* sp. might be confused on the island is *Phaeoceros carolinianus*, both having stick-like capsules. However, the diagnostic black colour of the spores of *Anthoceros* (×20 lens) easily distinguishes it from *Phaeoceros carolinianus*, which produces yellow spores. Sterile plants of *Anthoceros* can be recognised by presence of large, irregular cavities throughout the thallus (cross-section under the microscope) and by the more strongly crisped, ascending thallus margins. The thallus of *P. carolinianus* is solid, flat, and the margins only moderately undulate.

Habitats Open, loamy soil in garden flower-beds at Mount Pleasant, where it was freely fertile with sporulating capsules in October. In the same month, on a track-side near Botley's Lay, the remains of old and dry plants of an *Anthoceros* were found on loamy soil at base of a small rock scar, with a single capsule that still enclosed a few of the diagnostic black spores. It is not known whether the Mount Pleasant and the decayed Botley's Lay plants are the same species.

Status and distribution *Anthoceros* may be scarce and local on the island, or probably ephemeral, and it is certainly much more restricted than the common *Phaeoceros carolinianus*. *Range*: The sub-cosmopolitan genus *Anthoceros* is very poorly known in tropical areas, especially in Africa, and the identities of the St Helena plants are not yet known. A different species of *Anthoceros* (*A. cristatus*) occurs on Ascension Island.

LEFT **Non-fertile thalli** RIGHT **Thalli with sporophytes** INSET **Dehisced sporophytes showing yellow spores**

Phaeoceros carolinianus

Notothyladaceae

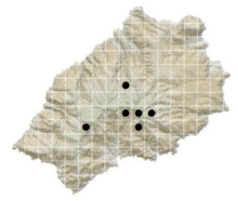

Description Thallus bright or dull green, smooth on the upper surface, 4–7 cells thick in cross-section, variably incised or divided into short lobes a few mm wide that are sometimes expanded at the apex; thallus solid, lacking internal cavities. Each thallus cell with a single, very large chloroplast. Monoicous. Antheridia in cavities in the upper surface of the thallus. A narrowly cylindrical, smooth involucre grows from the thallus surface and protects the developing capsule, the green capsule eventually extending through it and becoming linear, stick-like, up to 4 cm long, finally darkening and splitting from the apex to release the yellow spores.

Recognition *P. carolinianus* is most easily recognised by the flat, smooth thallus and the yellow spores produced from the stick-like capsules. The spores of *Anthoceros* are black, and those of *Dendroceros*, green. When sterile, the smooth thallus with only weakly undulate margins and the absence of internal cavities (cross section under the microscope) distinguishes it from *Anthoceros*.

Habitats Forming flat sub-rosettes or irregular patches (that are sometimes extensive) on shaded, moist soil and rocks, by paths and streams, on steep, vegetated roadside banks, cut earth steps, sometimes on wet mud in seepage areas.

Status and distribution Widespread on the island at middle and upper altitudes, often abundantly fertile. *Range*: *P. carolinianus* is a very widespread, cosmopolitan species.

LEFT **Thalli with abundant domed antheridial (male) chambers** TOP RIGHT **Non-fertile thalli** BOTTOM RIGHT **Young sporophytes**

Dendroceros adglutinatus

Dendrocerotaceae

Synonym *Monoclea adglutinata*

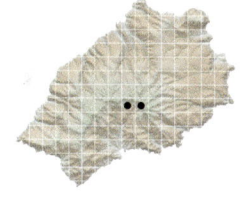

Description A small, thallose species, forming dark green patches of intricately overlapping thalli. Thallus usually profusely branched (the thallus repeatedly divided in a Y-shaped fashion into two equally-sized branches, which themselves are similarly divided). Thallus composed of a broad midrib 5–8 cells thick and wings 1 cell thick, the margins strongly undulate-crisped or sometimes weakly undulate, the ultimate lobes often weakly notched. Laminal cells each with a single, very large chloroplast. Monoicous. Antheridial cavities, each covered by a hemispherical dome, scattered over the thallus, often in dense clusters on the branches. A narrowly cylindrical, scaly, involucre grows from the thallus surface and protects the developing capsule, the green capsule eventually extending through it and becoming stick-like, up to 15 mm long; spores green at the time of release.

Recognition When fertile, the stick-like capsules of *D. adglutinatus* differentiate it from all other thallose species on the island, except for *Anthoceros* sp. and *Phaeoceros carolinianus* which, however, are found exclusively on rocks and soil and their spores are not green. The only other thallose species that may grow epiphytically on very moist bark and rotting wood are species of *Riccardia*, which are very different in growth form, do not have stick-like capsules and the cells have numerous small chloroplasts.

Habitats An epiphytic species occurring in humid locations on the moderately shaded branches of native trees and shrubs, including whitewood, he cabbage tree and black cabbage tree. It prefers smooth, young bark, and on black cabbage trees mostly occurs on the ultimate branches, especially near the terminal leaf rosettes. It is also frequent on the decaying petioles of tree fern and the branches of Norfolk Island pine, and occasionally on other, mostly alien trees and shrubs, including bilberry tree.

Status and distribution Endemic. Found only in Diana's Peak National Park at high altitudes, where it is relatively widespread, although sometimes in small quantities.

LEFT **Plants from near Cuckold's Point** RIGHT **Plants from the Depot**

Sphagnum helenicum

Sphagnaceae

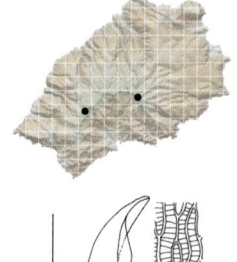

1 mm   leaf cells

branch leaf

Description Plants hummock- or mat-forming, green, or yellowish- to brownish-green (rarely reddish); stems weak, with short, spreading or ascending branches arising singly along the stem (occasionally in pairs), and short apical branches forming a compact head. In plants found at the Depot, the branches are usually curved and tapered to a sharp point ('cow-horn'); those of the Cuckold's Peak plants tend to be straighter, with more distant leaves. Branch leaves ovate, straight or curved, broad at base and tapering to a narrow or wide apex, lacking a nerve, comprising a network of narrow, green, elongate cells surrounding large, transparent cells that are usually reinforced with fibrils (fibre-like wall thickenings). Fertile plants not seen.

Recognition Species of the genus *Sphagnum* are morphologically highly distinct, and cannot be mistaken for any other bryophyte. Among the unique features of the genus are the branches that arise in bunches along the stem (but mostly singly in *S. helenicum*), and the cell structure of the leaves. The pattern of large, elongate, transparent, fibrillose leaf cells can be easily seen with a ×20 hand lens.

Habitats At the Depot, on a seasonally wet, steep, rocky, south-facing slope, where it occurs as scattered hummocks over a very small area, often associated with buck's-horn and tussocks of New Zealand flax. Near Cuckold's Point, in low, almost permanently submerged mats in a constantly wet hollow along a grassy path, with tiny population outliers in wet ground nearby.

Status and distribution Endemic; endangered at both of its two known locations in St Helena. Plants from the two locations differ in habit and growth form, perhaps habitat-induced, but could perhaps be recognised as varieties. However, the 19th century material collected by Burchell and others appears to be intermediate in character, and further work is required before the taxonomic status of the forms can be established.

TOP LEFT **Plants with sporophytes in all stages of maturity** BOTTOM LEFT **Male inflorescences** RIGHT **Mature and immature capsules**

Entosthodon sp.

Funariaceae

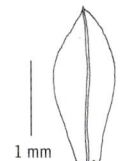

1 mm

Description Plants scattered or in tufts; leafy shoots erect, 0.5–1 cm high, branched at the base; leaves crowded in the upper shoot, forming an apical sub-rosette. Upper leaves concave, 1.7–2.5 mm long, nerve reaching apex, margin toothed. Leaf cells large, 40–60 × 20–30 μm in mid-leaf, the marginal cells longer and narrower and forming a distinct or indistinct border. Nerve and margins sometimes yellowish. Autoicous, capsules mature in October and November. Male branches with a cluster of antheridia in the rosette; in female shoots, seta long, capsule symmetrical, broadly pear-shaped with a small, flattened operculum (lid) and thin calyptra with a long, oblique rostrum.

Recognition The distinctive pear-shaped capsules gradually tapered at the base, and the small, flattened operculum lacking a rostrum distinguish *Entosthodon* from species of *Physcomitrium*, which have nearly globose, ovoid, or slightly pear-shaped capsules more abruptly tapering at the base, and a conical or rostrate operculum. Both genera can be recognised by the broad, large-celled leaves having a distinct margin of narrow cells. The leaves of *Entosthodon* may, in general, be smaller, have less sharply acute apices and rather shorter marginal cells than those of *Physcomitrium*, but they are variable, and the species cannot be safely identified in the absence of capsules, which appear to be seasonal.

Habitats A plant of open sites or partial shade, growing on bare soil in small, discrete tufts, or as scattered plants intermixed with other bryophytes. It has been found on roadside banks, on a steep rock/earth track-side cutting, on a cliff ledge and on open soil on a steep path.

Status and distribution Apparently endemic. Recorded from several locations along the Central Ridge, including on the cliffs of High Peak; also known near Sandy Bay village and by Spider Sprint path in Diana's Peak National Park.

LEFT **Mat in typical shaded habitat** RIGHT **Fertile shoots with sporophytes**

Fissidens chioneurus

Fissidentaceae

Synonym *Fissidens helenicus*

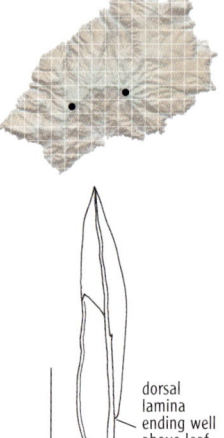

Description A large *Fissidens*, mid-green or sometimes yellowish-green, typically forming tufts or mats with branched or unbranched, frond-like shoots to 3–4 cm long, and individual branches to 2.5 mm wide. Leaves to 2 mm long, up to 20 or more pairs on a frond, somewhat crisped when dry, acute, the sheathing lamina with an intramarginal border for most of its length, the dorsal and apical laminae unbordered, and the dorsal lamina ending well above the leaf base. Leaf cells each with many papillae, obscuring the cell walls (microscope). Autoicous. Male inflorescences bud-like, in the leaf axils. Female inflorescence at the shoot apex, seta 5–7 mm long, capsule 1 mm long.

Recognition Although the large size of well-grown *F. chioneurus* is a good field character, *F. elegans* can be as large, and resemble it in the field, but its leaves are unbordered. *F. chioneurus* is the only *Fissidens* on the island with the combined characters of multi-papillose leaf cells, leaves with an intramarginal border restricted to the sheathing laminae, and the dorsal lamina ending well above the leaf base.

Habitats Grows in moderately to deeply shaded habitats, mostly on damp, rock/earth banks, or on steep to vertical rock scars; less frequently on tree bases, rotting trunks of trees and tree ferns, and rarely on decaying vegetation.

Status and distribution An endemic species, confined to the highest parts of the island. It is frequent in Diana's Peak National Park, with an outlier at the Depot, where it seems to be rare; perhaps to be found in remnant native vegetation elsewhere along the Central Ridge.

dorsal
lamina
ending well
above leaf
base

0.5 mm

Various forms BOTTOM RIGHT **Showing bordered laminae**

Fissidens curvatus subsp. *helenicus*

Fissidentaceae

Description A small *Fissidens* with unbranched shoots to 3 mm long and 1 mm wide. Leaves 1–1.5 mm long, shortly pointed, the nerve reaching the apex, and all laminae strongly bordered, the borders reaching the leaf apex or ceasing just below it; the basal part of the dorsal lamina consisting solely of border tissue. Leaf cells smooth. Autoicous. Fertile and sterile shoots differ: the female shoots resembling the frond-like vegetative shoots but shorter and having fewer, longer leaves. Male plants bud-like (hardly detectable in the field), at the base of female and vegetative fronds.

Recognition On St Helena, this species is rather uniform, and is the only *Fissidens* in which all the leaf laminae are strongly and continuously bordered. The border can just be discerned in the field with a hand lens, but microscopic examination is necessary for certain identification of *F. curvatus* subsp. *helenicus* and the other small *Fissidens* on the island, including *F. serratus, F. taylorii* and *F. tenellus*. Elsewhere in its world range, *F. curvatus* is quite variable in leaf shape and in the development of the leaf borders.

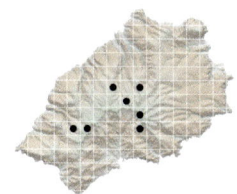

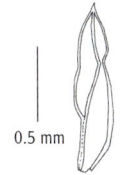

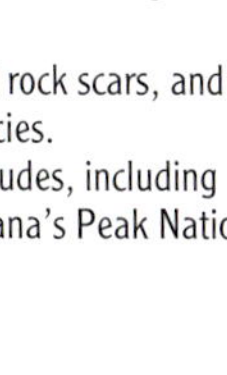

0.5 mm

Habitats Grows on moist, shaded soil banks (clay, silt, gritty or humus-rich), on thin soil overlying rocks or directly on rock, typically scattered or small, loose groups of shoots. Found on stream banks, vegetated roadside banks, on moist cliffs and rock scars, and by waterfalls. It is often found with *F. translucens*, and sometimes with other *Fissidens* species.

Status and distribution An endemic subspecies. Fairly widespread, mostly at middle altitudes, including at Bishop's Bridge, Blue Hill village, Alarm Forest, Wash House Gut; also by streams in Diana's Peak National Park and on cliffs at High Peak. *Range*: Worldwide *F. curvatus* is sub-cosmopolitan.

LEFT & RIGHT Shoots growing at base of *Jensenia spinosa* INSET Close-up of portion of leaf showing thick-walled cells

Fissidens porrectus

Fissidentaceae

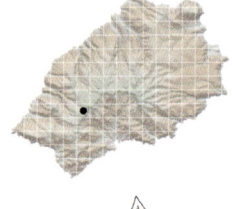

Description A rather small species of *Fissidens*, unbranched or branched, pinnate, the frond-like shoots 4–6 mm long, and leaves rather widely spaced along the stem (but can be closely overlapping in other parts of its range). Leaves 0.5–1.1 mm long, the nerve ending 2–9 cells below the sharp apex, all laminae unbordered (or sheathing lamina weakly bordered), leaf margins smooth but often somewhat undulate (wavy), and two cells thick in part. Leaf cells smooth, often with very thick walls and rounded cell cavities. Fertile plants not known from St Helena.

Recognition The St Helena plants are characterised by the unbordered (or weakly bordered) leaves, and the smooth, clearly defined leaf cells with very thickened walls and rounded cell cavities (microscope). The widely spaced leaves also lend a distinctive appearance to the shoot, but this may be induced by the habitat and is perhaps not a constant character. *F. porrectus* is a variable species, and elsewhere in its range possesses features not seen in the St Helena plants, including leaves with distinct dark borders. If further material is found on St Helena, it will be interesting to see if the species is uniform or variable on the island.

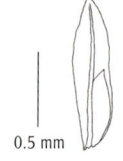

0.5 mm

Habitats Found as scattered shoots in a tuft of *Jensenia spinosa* on sheltered, moist rock on the south-facing cliffs of High Peak.

Status and distribution On St Helena, it is currently known only from the single location, where a very small quantity was discovered in 2009. Its absence from the extensive collections of *Fissidens* made in 2005 suggest it is rare on the island. *Range*: Widely distributed in west and east Africa.

LEFT **In situ at the Bellstone** RIGHT **Herbarium plants with some colour lost**

Fissidens pygmaeus

Fissidentaceae

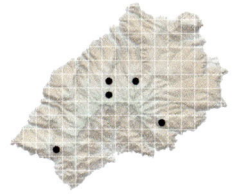

Description Plants small, yellowish-green, growing in loose mats or as scattered shoots. Shoots frond-like, unbranched, the vegetative and fertile shoots differing in form. Vegetative shoots to 8 mm high and 0.2–0.7 mm wide; leaves somewhat shrunken when dry, fairly uniform in size, 0.3–0.6 mm long, distant on the stem, the upper part of the nerve and the upper part of the leaf often conspicuously and sharply angled. Dioicous or autoicous. Fertile shoots shorter, but the leaves markedly increasing in size from the base of the shoot; basal leaves tiny, the upper part somewhat bent to one side, apical leaves to 1.2 mm long and straighter, perichaetial leaves to 1.6 mm long. The sheathing laminae are variably bordered. Capsules have not been reported on St Helena collections, although plants with archegonia have been seen.

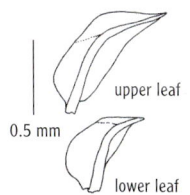

upper leaf

0.5 mm

lower leaf

Recognition Although a small species, *F. pygmaeus* can be recognised in the field by the curved or sharply angled upper part of the leaf and nerve, and by the different morphology of the fertile and vegetative shoots.

Habitats A plant of bare soil in partly shaded locations, found on banks by tracks, at the edge of woodland, on earthy roadside banks and at the base of a rock outcrop in pasture. Reported associates include *Dicranella proscripta*, *Cephaloziella* spp., *Bryum sauteri* and *Cladonia* spp. (lichens).

Status and distribution This species has been recorded in five locations: near Botley's Lay, at Cason's, the Bellstone, near Bishop's Bridge and Napoleon's Tomb. *Range*: Known only from South Africa, Lesotho, Zimbabwe and Tanzania, where it is apparently local and rare.

LEFT **Dry shoots** CENTRE **Moist shoots** RIGHT **Sporophytes**

Fissidens reimersii

Fissidentaceae

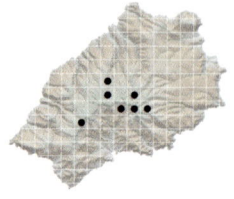

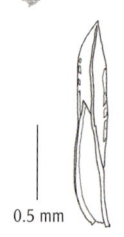

0.5 mm

Description A rather small species, with unbranched frond-like shoots 2–7 mm long, each branch up to 1.7 mm wide and with up to 16 pairs of leaves. Leaves 1.3–1.8 mm long, dark to glaucous-green, contorted when dry, nerve almost reaching apex, the sheathing lamina with a marginal border throughout its length, the dorsal and apical laminae with an intramarginal border that is usually discontinuous. Dorsal lamina ending well above the leaf base. Laminal cells with many conspicuous papillae that often obscure the cell walls. Frequently fertile: seta 2–3 mm long, capsule 0.6 mm long, with red, conical lid.

Recognition *F. reimersii* can be recognised in the field by its rather dark colour, and leaves with clearly defined, translucent borders. The combined characters of borders on all laminae, dorsal lamina ending well above the leaf base, and leaf cells each with many papillae is unique in the genus.

Habitats A plant of freely draining soils and porous rock in moderately to densely shaded habitats, forming dense or loose colonies, or found as scattered shoots amongst other bryophytes: open soil on roadside and stream banks, rock scars, and earthy/stony soil shaded by shrubs or ferns. A wide range of associates includes *Cephaloziella* sp., *Dicranella proscripta* and *Philonotis helenica*.

Status and distribution Endemic. *F. reimersii* is one of the most widespread species of *Fissidens* on the island, found at middle to high altitudes. It is frequent in Diana's Peak National Park and elsewhere along the Central Ridge; also in many other places in the central parts of the island including upper Deep Valley, Bishop's Bridge, Swampy Gut, Cason's and Hutt's Gate.

LEFT & RIGHT **Growing with *Fissidens curvatus* subsp. *helenicus***

Fissidens serratus

Fissidentaceae

Description Shoots small, frond-like, 2–2.5 mm long × 1.2 mm wide. Vegetative fronds each with up to 10 pairs of leaves, slightly crisped when dry, 0.8–0.9 mm long, margins serrate, unbordered; perichaetial leaves to 1.3 mm long, with a weak, short border on the sheathing laminae, margins serrate. Dorsal laminae slightly rounded below, reaching the insertion, not decurrent. Nerve extending almost to the apex on St Helena plants. Laminal cells bulging. Seta 2.5 mm long, smooth; capsule ovoid, 0.4 mm long.

Recognition This species is recognised by its small size, bulging leaf cells, unbordered stem leaves, perichaetial leaves unbordered or weakly bordered, and the usually conspicuously denticulate-serrate margins of the laminae. Microscopic examination is required. *F. serratus* is variable in the shape of the plants, width of the leaves and leaf tips, extension of the nerve, laminal cell size and serration of leaf margins.

Habitats *F. serratus* is found on damp, rather deeply shaded rocks, typically under a canopy of trees or shrubs: elsewhere in its range, recorded from soil, trees and rotten wood.

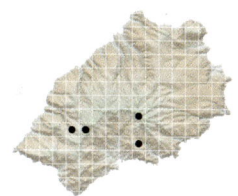

0.5 mm

Status and distribution This species has been found in a wooded valley near Blue Hill village, the south-east slopes below Cuckold's Point, near the waterfall in Wash House Gut and below a sheltered cliff ledge on High Peak. *Range*: Pantropical.

Further species A few shoots of a similar species, *F. tenellus*, have been found on damp rocks near Jockie's Gut, in Diana's Peak National Park. It is difficult to distinguish from *F. serratus*, but differs in having an intramarginal border on the sheathing laminae of the upper leaves of the fertile female shoots, and the leaf margins tend to be only weakly serrate. *Range:* It is known also from Australasia, and from islands in the western Indian Ocean.

LEFT & CENTRE **Moist shoots** RIGHT **Dry shoots**

Fissidens taxifolius

Fissidentaceae

Description A rather large species with unbranched, frond-like shoots up to 2 cm long × 2 mm wide, each with up to 15 pairs of leaves. Leaves 2 mm long, crisped when dry, oblong-ovate and tapering rather abruptly to an obtuse apex, the leaf margins minutely and regularly serrate (toothed), the nerve excurrent in a rather long, sharp point, the leaf cells strongly bulging.

Recognition Because of its relatively large size, this species is unlikely to be mistaken for any other *Fissidens* on the island. It is easily recognised by the rather large, wide, unbordered leaves abruptly tapered to the apex, and the nerve conspicuously excurrent in a sharp point. Microscopically, the finely toothed leaf margins and bulging cells are key features.

Habitats On clay-loam soils in open or shaded places, including on a low path-side bank, on the woodland floor, and on thin soil on a shaded concrete block. Reported associates include *F. translucens, Bryum sauteri, Dicranella proscripta* and *Cephaloziella* sp.

Status and distribution Apparently scarce on St Helena, and currently known from only three locations in a single 1 km grid square: St Paul's Cathedral grounds, Plantation Wood and by the path to the Boer Cemetery. *Range*: Sub-cosmopolitan: Europe, Macaronesia, northern and South Africa, Australasia, North and South America.

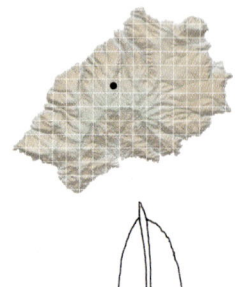

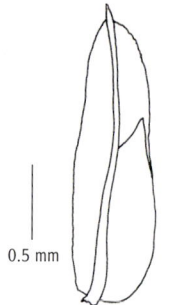

0.5 mm

LEFT Sterile shoots RIGHT Fertile shoots with sporophytes

Fissidens taylorii

Fissidentaceae

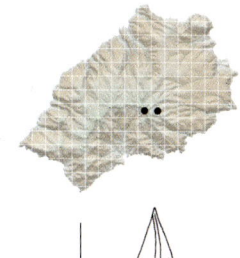

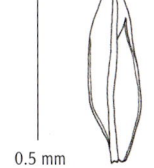

0.5 mm

Description A very small species, the vegetative and fertile shoots differing in form. Vegetative (non fertile) shoots frond-like, 0.8–1.2 mm long; leaves 0.4–0.65 mm long, hardly crisped when dry, distant, up to 8 pairs in a frond, margins unbordered, or weakly bordered in the sheathing lamina and the middle part of the dorsal lamina; sheathing lamina about half the leaf length; nerve ending near or in the apex; cells in mid-leaf 9–15 µm long. Autoicous. Female plants also frond-like, unbranched, 0.8–1.1 mm long; leaves in 3–4 pairs; border in upper leaves variable; perichaetial leaves to 1.5 mm long. Male plants tiny, bud-like, 0.4 mm long, at the base of the female plants. Seta 2–4.5 mm long, smooth; capsules sub-erect, 0.3–0.5 mm long.

Recognition *F. taylorii* cannot be certainly identified in the field with a ×20 hand lens. Under the microscope, the differing vegetative and fertile shoots, with the basal, bud-like male plants (careful examination required), the few-leaved female shoots with long perichaetial leaves, and the many-leaved vegetative shoots distinguish *F. taylorii* from all other tiny species of *Fissidens* on the island, except *F. tenellus* (which, however, differs in its papillose cells). The leaves of *F. taylorii* vary from unbordered to bordered on all laminae; borders can end well below the leaf tip or almost reach it.

Habitats Occurring as loose mats or scattered shoots, on rocks in guts subject to spray and/or seasonal inundation, associated with *F. translucens* or *F. curvatus* subsp. *helenicus*.

Status and distribution *F. taylorii* is apparently local on St Helena, and currently known from only two small areas, in the Jockie's Gut area of Diana's Peak National Park and in upper Deep Valley. *Range*: Australia, New Zealand, southern United States, and Central and South America.

TOP LEFT & TOP RIGHT **Mature and dehisced sporophytes** BOTTOM LEFT **Moist shoots** BOTTOM RIGHT **Dry shoots**

Fissidens translucens

Fissidentaceae

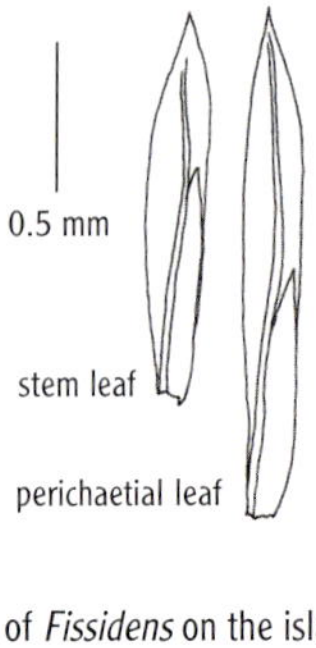

Description A small species, with unbranched, frond-like shoots 3–4 mm high and 1–1.5 mm wide, each with few pairs of leaves; stem with low, swollen nodules. Leaves variable, to 2.2 mm long, often long and very narrow (especially perichaetial leaves of fertile shoots which can be up to 9 times as long as wide), acute, the nerve ceasing below the apex, all laminae unbordered, margins very weakly toothed. Leaf cells smooth, often becoming large and lacking chlorophyll adjacent to the nerve, so that the nerve appears to have a translucent zone of cells each side of it. Dioicous or autoicous.

Recognition Although a small species, *F. translucens* can often be recognised in the field, with experience, by the long, narrow leaves and translucent zone of cells alongside the nerve (×20 lens). However, the zone of translucent cells is not always apparent, and microscopic examination is required for certain identification. Other small species of *Fissidens*, including *F. serratus* and *F. taylorii*, usually have shorter leaves. Microscopically, the smooth cells, the absence of leaf borders, and the presence of swollen nodules on the stem are key characters.

Habitats *F. translucens* is found in moist, very shaded environments, growing in diffuse, or sometimes dense, populations, usually on rock or stones (including boulders by streams, rock outcrops in valley bottoms and by waterfalls), less frequently on soil and rarely on tree stumps. It is often found in association with other *Fissidens* species, including *F. curvatus* subsp. *helenicus* and *F. taylorii*.

Status and distribution Endemic. *F. translucens* is one of the most widespread species of *Fissidens* on the island. It is frequent in the lower parts of Diana's Peak National Park and also at lower altitudes.

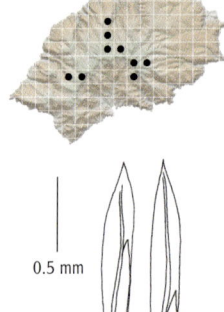

0.5 mm

stem leaf

perichaetial leaf

LEFT **Dry shoots** CENTRE & RIGHT **Moist shoots**

Ceratodon purpureus

Ditrichaceae

Description Plants forming tufts or mats, variable in colour, yellowish-green or deep green to reddish; shoots 1–3.5 cm high, unbranched. Leaves 1–1.5 mm long, spreading when moist, appressed and twisted when dry, channelled, margins narrowly recurved from base to near apex, often with a few blunt teeth near apex, nerve stout, sometimes very shortly excurrent. Leaf cells rectangular below, square to irregularly hexagonal above (sometimes the cell cavity with rather rounded corners), 9–12 µm, smooth, clear. Dioicous. Female plants with deep purple-red setae, the capsule ovoid, erect when young, becoming inclined then horizontal at maturity.

Recognition A variable and rather anonymous moss, which can make it difficult to identify in the field. However, characteristic features are the rather spiky appearance of the shoots (when moist), leaf margins recurved for most of their length and, under the microscope, the non-papillose cells. Fertile plants are easily recognised by the bright purple-red setae bearing an erect capsule that later becomes horizontal, but fruiting plants have not yet been reported from the island.

0.5 mm

Habitats Found in a wide range of habitats: wall-tops, open soil patches in rough pasture, earthy banks and gritty soil along paths. Elsewhere in its range, it is also found on heathland, fire sites, rocks, old tree stumps and fence posts.

Status and distribution Widespread, though apparently fairly local on the island, with records from Hutt's Gate, Level Wood Reservoir, Napoleon's Tomb, High Knoll Fort, Cleugh's Plain and White Point. *Range*: Sub-cosmopolitan, with a number of subspecies described, based on the characters of the capsule and peristome.

LEFT & CENTRE Fertile shoots with *Ceratodon purpureus* RIGHT Shoot showing capsule and long perichaetial leaves

Pleuridium acuminatum

Ditrichaceae

Description Plants very small, green or yellowish-green, shoots erect, 3–5 mm high, usually unbranched. Leaves narrow, short in the lower part of the shoot, longer above; the upper and perichaetial leaves to 1.8 mm long, the fine point consisting mainly of nerve; the nerve below broad but not sharply defined; leaf margins lacking teeth. Leaf cells narrowly rectangular to sub-hexagonal. Antheridia in the axils of perichaetial leaves. Capsule ovoid and bluntly pointed, on a very short seta, immersed in, and overtopped by, the cluster of long perichaetial leaves.

Recognition This small species is inconspicuous in the field when sterile and/or occurring as scattered shoots amongst other mosses, although the hair-like leaf apices may alert the observer to its presence. However, it usually occurs in small clumps of shoots, and is often abundantly fertile, when it can be recognised by the short, immersed capsules, viewed from above looking somewhat like rounded grains sitting in the tuft.

Habitats A plant generally of freely draining soils in open habitats. At White Point, in rough upland pasture where the grass cover is patchy, growing with *Campylopus* sp. and *Ceratodon purpureus*. At Middle Point, Longwood, on small areas of bare soil in an open, herb-rich, scrubby area, associated with *Campylopus introflexus* and *Trichostomum brachydontium*.

Status and distribution There are currently only two records of this inconspicuous species, but it may prove to be quite widespread at mid-altitudes on the island. *Range*: Europe, Macaronesia, Africa, Caucasus, China and North America.

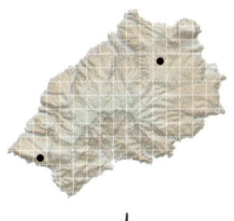

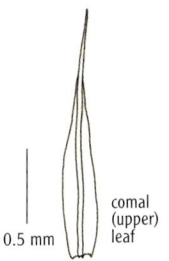

0.5 mm

comal (upper) leaf

LEFT **Tuft on woodland floor** RIGHT **Examples of leaf form**

Campylopus arctuatus

Dicranaceae

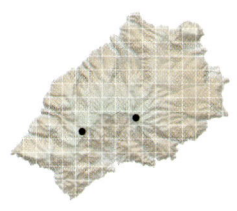

Synonym *Campylopus helenicus*

Description Plants mid- to olive-green, forming loose tufts or cushions; shoots to 7 cm high, tomentose (sometimes to near shoot apex). Leaves 5–6 mm long, narrow, weakly curved to sickle-shaped, very gradually tapering to a narrow, smooth or weakly toothed point; nerve very broad, $^1/_3$–$^1/_2$ width of leaf base. Basal cells of leaf rectangular, evenly thick-walled (or somewhat unevenly), the alar cells differentiated into large-celled, reddish, ear-like projections at the leaf base; cells in mid and upper leaf short and small, rhomboidal.

Recognition *C. arcuatus* is characterised by its long, narrow, curved leaves with a (usually) nearly smooth apex, and notably small upper leaf cells, these characters separating it from *C. flexuosus*, some forms of which may also differ in having deciduous branchlets at the shoot tip

Habitats *C. arcuatus* grows in open or shaded places in a wide range of habitats including on the trunks and branches of tree fern, black cabbage trees and Norfolk Island pine, and on peaty banks, rotting vegetation and cliff ledges.

Status and distribution *C. arcuatus* is confined to the highest parts of the island, mostly to Diana's Peak National Park. *Range*: Ascension Island, Malawi, and the Indian Ocean islands, but probably overlooked in continental Africa.

Further species Another St Helena species, *C. thwaitesii,* can resemble *C. arcuatus*, but is described as differing in its strongly toothed leaf apex, in the unevenly thickened basal leaf cells (the cell walls appear knobbly), and in the nerve cross-section. Unfortunately, there are some inconsistencies in published descriptions of the two species, and therefore only a tentative identification to species is possible for some collections at the present time. *C. thwaitesii* is found in the same habitats as *C. arcuatus*, and also appears to be confined to the highest parts of the island. It is recorded from southern Africa, Sri Lanka, Sumatra to New Guinea and Brazil.

LEFT Shoots with long hair points RIGHT Shoots with shorter hair points

Campylopus pilifer

Dicranaceae

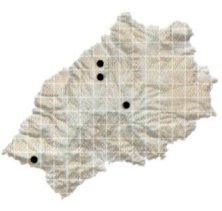

Description Plants dark olive-green to yellowish-green above, lower parts
blackish-green; shoots 1–3 cm high, of uniform width throughout (lacking
thickened nodes), tomentose below but sometimes sparsely so. Leaves
appressed when dry, spreading-erect when moist, 2.5–5 mm long. Leaf nerve
very broad, $^1/_2$ the leaf width, excurrent in a hyaline, strongly toothed hair-
point of variable length that is usually almost straight when both dry and
moist; the back of the leaf nerve with prominent ridges (lamellae). Lower part
of leaf with a large area of narrowly rectangular hyaline cells that extend
higher at the margins than by the nerve, and is sharply delimited from the
small chlorophyllose cells above.

Recognition The hyaline zone of cells in the leaf, and the hyaline, strongly toothed leaf hair-point will separate
C. pilifer from all other species on St Helena except *C. introflexus* in which, however, the hair-point is strongly
reflexed in the dry plant. The shoots of *C. pilifer* are usually darker in colour, and are uniformly thick, whilst
those of *C. introflexus* are often paler and thickened at the nodes. However, for certain identification, microscopic
examination is desirable, the two species differing in the anatomy of the leaf nerve as seen in cross-section (the
lamellae are 3–4 cells high in the upper part of the leaf in *C. pilifer*, but only 1–2 cells high in *C. introflexus*).

Habitats On St Helena, recorded from dry rocks in open locations, and on a roadside stone wall. Elsewhere in its
range, it is also found on freely draining soils on heaths and dunes, and in open scrubland.

Status and distribution As yet, reported only at White Point, High Knoll Fort and nearby, and on a wall at Hutt's
Gate. *Range*: Coastal regions of western Europe, Macaronesia, Africa, India to south-east Asia, southern North
America and South America.

A note on *Campylopus* – it should be noted that the identification of specimens in this 'difficult' genus may be
particularly problematic, since not only does identification usually depend on examining cellular structures, but the
species also display a bewildering array of forms. Four species are described in this book in order to show the sorts
of characteristics that are found in the genus. Several other forms (or species) are known from the island, but not
yet identified.

LEFT **Fertile shoots in situ on roadside bank** CENTRE **Herbarium material with dehisced sporophytes** RIGHT **Leaf**

Dicranella proscripta

Dicranaceae

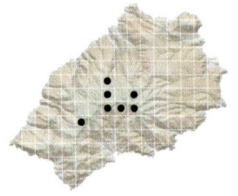

Description Plants yellowish-green to deep green, forming loose tufts or more extensive mats, or often found as scattered plants in mixed bryophyte communities. Shoots erect, usually 2–4 cm high, simple or with occasional branches. Leaves 3–6 mm long, the base broad, clasping the stem or the base of the adjacent leaf, quickly contracting to a long, narrow limb composed mostly of nerve above, and spreading-flexuose from the shoot; the apical part of the perichaetial leaves being conspicuously long, narrow and flexuose. Seta pale or dark, 6–10 mm long, flexuose; capsule ovoid or ovoid-cylindrical, 1–1.2 mm long, operculum with a long, sharp rostrum; spores papillose but not coarsely warty.

Recognition Well-grown plants of *D. proscripta* are easily recognised in the field by the hair-like, strongly flexuose leaves, rapidly contracted above the broad base, and when fertile by the long, flexuose seta and long-rostrate capsule. The leaves are similar in shape to some other *Dicranella* spp., and although plants of *D. proscripta* are generally much larger, stunted forms occur, and it is therefore unsafe to name small, non-fruiting plants.

Habitats Grows on a wide range of loamy, gritty, peaty or humus-rich soils on shaded earthy or rocky banks, on soil-capped rocks and in rock crevices. It is frequent on steep banks in cloud forest communities in Diana's Peak National Park, and also on steep, shaded roadside banks elsewhere.

Status and distribution Presumed endemic to St Helena and Ascension Island. It is widespread on St Helena at middle to upper altitudes, and frequent in Diana's Peak National Park. The genus is poorly known in Africa and elsewhere.

LEFT **Sporophytes with long, pointed calyptrae** CENTRE **Moist shoots** RIGHT **Dry shoots**

Chionoloma bombayensis

Pottiaceae

Synonyms *Pseudosymblepharis bombayensis, Trichostomum bombayense*

Description Plants variable, olive to golden-green, forming low mats of ascending shoots 1–2 cm high. Leaves 2–4 mm long, apex obtuse or acute; nerve excurrent in a short, abrupt point; margins plane, often somewhat undulate. Leaves strongly curled when dry, back of the nerve glossy; when moist, usually spreading-erect, the margins flat below and usually erect or partly incurved above. Basal cells of leaf hyaline, thin-walled, sometimes ascending higher along leaf margin than by the nerve (but variable); above that a small zone of clear, thick-walled cells, then very small, opaque, highly papillose cells in the rest of the leaf. Fertile plants common: seta 0.9–1.4 mm long, reddish; capsule narrowly cylindrical, 2 mm long; peristome teeth long and narrow, straight or slightly twisted.

Recognition On St Helena, this variable species is best differentiated in the field from similar species by the narrowly cylindrical capsules. Vegetatively, it can be separated from *Trichostomum brachydontium*, when moist, by the less spreading leaves with turned up margins above. The two species may grow close together in tufts or mats. Under the microscope, the sharper transition between the hyaline, basal and small, opaque, upper cells of the leaf should separate plants of *C. bombayensis* (the hyaline cells usually extend rather higher up the leaf margin than they do along the nerve). *T. brachydontium* is never epiphytic.

Habitats On open, freely draining soil, on walls, stone steps and a stone gate-post, and less frequently on tree branches. It often occurs in association with *Weissia* spp. and *Trichostomum* spp. on soil, and probably also occurs on natural rock outcrops on the island.

Status and distribution Fairly widespread and probably frequent on the island. *Range*: Very widespread in Africa and Asia.

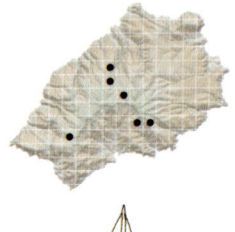

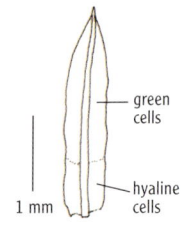

green cells

hyaline cells

1 mm

LEFT **Moist shoots** RIGHT **Dry shoots**

Didymodon sp.

Pottiaceae

Description Plants forming open or dense colonies closely attached to the
substrate, olive-green or brownish, dull; shoots erect, 2–3 mm high,
sometimes branched. Leaves crisped and somewhat appressed when dry,
widely spreading-recurved when moist, 1.2–1.7 mm long, v-shaped in cross-
section and keeled, tapering to a rather blunt or sometimes acute apex,
sometimes shortly excurrent; margins crenulate, narrowly but conspicuously
recurved from near base to near apex. Basal leaf cells elongate- or shortly
rounded-rectangular, grading to small upper cells with irregularly sized,
rounded cell cavities and a single papilla. Fertile plants not observed.

Recognition Field characters of this species include the small size, the v-shaped,
spreading-recurved leaves with narrowly recurved margins and an often blunt
apex. Microscopically, the papillose upper cells further distinguish it. Because
so little material has been found and there are no capsules (which would help
to identify the genus), this plant cannot be positively identified, and is only
provisionally placed in *Didymodon*. However it could belong to another genus
of the Pottiaceae (e.g. *Gymnostomum*).

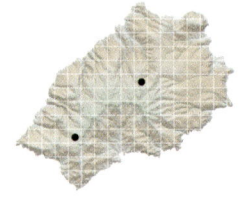

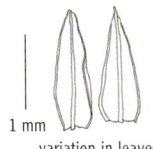

1 mm

variation in leaves

Habitats This plant has been found in small amounts growing on the mortar of a wall of a derelict building (the
Magazine) in open woodland near Hooper's Rock, and on the riser of a concrete step by Napoleon's Tomb.

Status and distribution Currently only two very small collections are known, both scrapings from the substrate.
It is possible that other species of *Didymodon* or related genera occur on the island, and careful observation of
low patches on soil, rocks or walls may reveal their presence. To detect other species, observers could look for
differences in, for example, the shape and orientation of the leaves, characters of the leaf margin and the size
and shape of the leaf cells.

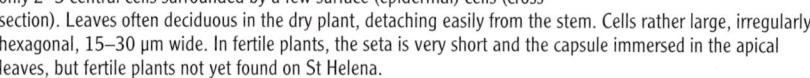

LEFT & TOP RIGHT **Moist shoots** BOTTOM RIGHT **Dry shoots**

Leptophascum leptophyllum

Pottiaceae

Synonyms *Tortula vectensis; Chenia leptophylla*

Description A small moss up to 8 mm high, in loose mats or as more scattered shoots, green, often brownish below, rather dark/dull when dry. Leaves broad, 1.5–2 mm long, widest above mid-leaf; when moist, rather soft in texture, widely spreading from the stem and somewhat recurved, with reflexed tips, margins irregularly denticulate above; when dry, the margins often strongly incurved and the leaves often appearing somewhat claw-like, but sometimes contorted and the margins not much incurved. Leaf nerve weak, consisting of only 2–3 central cells surrounded by a few surface (epidermal) cells (cross-section). Leaves often deciduous in the dry plant, detaching easily from the stem. Cells rather large, irregularly hexagonal, 15–30 μm wide. In fertile plants, the seta is very short and the capsule immersed in the apical leaves, but fertile plants not yet found on St Helena.

Recognition This species is quite easily recognised in the field, when dry, by the curved, crisped, or often claw-like appearance of the leaves. When moist, some species of *Bryum* may appear superficially similar, but *L. leptophyllum* can be differentiated by the recurved leaves with more strongly reflexed apices and larger leaf cells (×20 hand-lens). The few-celled, weak nerve (microscope) is diagnostic for this species on the island.

Habitats Mostly growing on freely draining, earthy or sandy soil, but sometimes on stone, in deeply shaded to fully exposed sites: on earthy banks, soil-covered tree roots, track-sides, in cultivated fields, on stony soil on graves, on stone steps and in flower pots.

Status and distribution *L. leptophyllum* is quite widespread on St Helena, thus far reported from Francis Plain, the Bellstone, St Paul's, Bishop's Bridge, Level Wood and St Matthew's Church at Hutt's Gate and a few other locations. *Range*: Widespread in scattered localities worldwide, including Ascension Island and several other remote islands.

LEFT & BOTTOM RIGHT **Moist shoots** TOP RIGHT **Dry shoots**

Pseudocrossidium crinitum

Pottiaceae

Synonym *Barbula crinita*

Description Plants forming loose tufts or cushions; leafy shoots erect, to 2 cm or more high, dull olive green, becoming darker or blackish below, appearing hoary when dry because of the long pale hair-points. Leaves appressed, spirally-twisted and somewhat contorted when dry, spreading-erect and channelled when moist. Leaf lamina to 2.5 mm long; margin rather broadly recurved, at least above; nerve sometimes reddish, broad in lower part of leaf, and excurrent in a long, hyaline, almost smooth hair-point up to 2 mm long. Cells in lower part of leaf, hyaline, narrowly rectangular, grading quite rapidly above to very small, nearly square cells, each cell having many forked papillae (making that part of the leaf look rather opaque). Fertile plants not known from the island.

Recognition The cushion-forming habit, relatively tall shoots and hoary appearance (especially when dry) are the first indications of this species. The tongue-shaped leaves with very long, silvery hair-points distinguish it from all other St Helena species except *Tortula muralis*. However, the latter is a much shorter plant, with leaves more spreading and a brighter green when moist, with a narrower nerve and more narrowly recurved leaf margins.

Habitats On the Haystack, on freely draining, gritty volcanic soil near rocky outcrops, growing in association with creeper and lichens; also on the dry, lower slopes of Sugar Loaf.

Status and distribution This species is known from only two locations on the island, and should be searched for in other areas on freely draining soils. *Range*: Northern and southern Africa, North and South America, and Australasia.

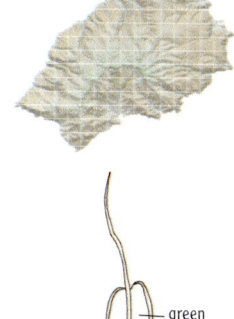

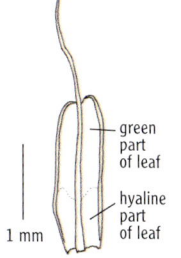

green part of leaf

hyaline part of leaf

1 mm

LEFT **Moist shoots** RIGHT **Dry shoots**

Tortula atrovirens

Pottiaceae

Synonym *Desmatodon atrovirens*

Description A small moss, green or yellowish-green when moist, dark brownish when dry, forming patches on open soil; shoots only a few mm high, the lower part sometimes buried in the soil leaving only 1 mm protruding above the soil surface. Leaves spirally-twisted and curved like a talon when dry, spreading when moist, concave, 1.2–1.5 mm long, margins recurved. Nerve broad, typically more broadened and thickened in upper half of the leaf, with prominent, enlarged, papillose cells on the upper side. Basal cells of leaf shortly rectangular, thin-walled, smooth; upper cells small, irregular, nearly square to sub-hexagonal, 10–12 µm wide, each with many papillae (except the 1–3 rows of marginal cells). Seta yellowish, up to 6 mm long; capsule erect, ovoid, 1 mm long; peristome teeth pale or whitish.

Recognition The small size, dark colour, spirally-twisted, talon-like dry leaves, and the nerve thickened above, will identify this species in the field. The few other species that occur in the dryland zone are very different in habit and form, and in the orientation of the dry leaves.

Habitats On open soil in sparsely vegetated or bare habitats: on a mud bank at the base of Rupert's Hill near the power station, and on dry soil on ledges and in rock crevices near the summit of Sugar Loaf, growing with *Bryum argenteum*.

Status and distribution Although there are only two records to date, *T. atrovirens* is likely to be found widely in the drier parts of the island. *Range*: Sub-cosmopolitan in temperate environments: Europe, the Middle East, Macaronesia, northern and southern Africa, North and southern South America and Australasia.

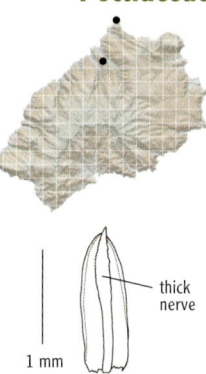

thick nerve

1 mm

underside of leaf

TOP LEFT **Moist shoots** BOTTOM LEFT **Dry shoots** RIGHT **Sporpohytes**

Tortula muralis var. *muralis*

Pottiaceae

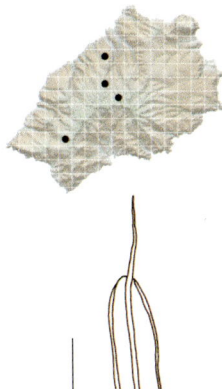

1 mm

Description Plants mat-forming or in small cushions, bright green when moist, darker below, appearing hoary when dry because of the long, pale leaf hair-points; leafy shoots 2–8 mm high. Leaves tongue-shaped, spirally-twisted, curved and appressed when dry, flat and widely spreading when moist, the lamina to 2.5 mm long; margins narrowly recurved, apex obtuse; nerve excurrent in a long, smooth, hyaline point up to 1.4 mm long. Cells in lower part of leaf, hyaline, rectangular, grading above to small, rather square cells, each with many papillae, obscuring the cell walls. Fertile plants frequent. Seta yellowish, up to 16 mm long; capsule narrowly cylindrical, 2–3 mm long; operculum narrow, 1–1.5 mm long. Var. *aestiva*, a form very different in appearance, has narrowly tongue-shaped leaves and a short, yellowish-green hair-point.

Recognition *T. muralis* var. *muralis* is easily recognised in the field by the short, mat-forming shoots, the tongue-shaped, hair-pointed leaves, narrow capsules and by its frequent occurrence on man-made structures. The only other species on the island with obtuse leaves and a long hyaline leaf point is *Pseudocrossidium crinitum* (for the differences, see account, p. 89). *T. muralis* var. *aestiva* differs from var. *muralis* in its narrower leaf and short hair-point.

Habitats Var. *muralis* has been found only on man-made structures on the island, including boundary walls, walls of old buildings and other stone structures, but is sometimes recorded from hard soil elsewhere in its range. Var. *aestiva* has been found only on a wall at St Paul's, near populations of var. *muralis*, and may be generally rare.

Status and distribution Var. *muralis* is probably widespread at mid-altitudes (though sparse); var. *aestiva* is likely to be rare. *Range*: A sub-cosmopolitan species found throughout the temperate and cool regions of the northern and southern hemispheres.

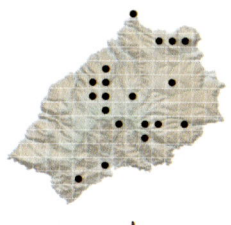

LEFT **Moist shoots** RIGHT **Dry shoots**

Trichostomum brachydontium

Pottiaceae

Description Plants forming yellowish- to mid-green, compact cushions or patches; shoots erect, to 3 cm high, sparsely branched. Leaves 2–4 mm long, widely spreading when moist, variable in shape, apex blunt or acute, margins flat; nerve excurrent in a sharp point of variable length. Basal cells of leaf rectangular, thin-walled, with a gradual transition to very small, opaque, highly papillose cells above. Dry leaves usually only moderately curled, not crisped or contorted, the pale, glossy back of the nerve apparent in the curled leaf; moist leaves widely spreading. Fertile plants occasional: seta to 10 mm long, capsule ovoid, 1 mm long.

Recognition Similar in size and general appearance to *Chionoloma bombayensis*, but the moist leaves are generally flatter and the margins not incurved, and in the dry plant, the leaves normally appear to be less strongly curled. However both species are variable and should be examined microscopically to compare the gradual transition between the basal hyaline and upper opaque cells in *T. brachydontium*, and the well-demarcated basal zone of hyaline cells that often ascend the margin in *C. bombayensis*. The long capsule of the latter species easily distinguishes it from *T. brachydontium*.

Habitats On earth banks, rocky soil among shrubs, soil-capping on walls and on mortar, on paths and verges, in crevices of rocky outcrops, at the edge of cultivated fields, occasionally in flower beds. *T. brachydontium* is one of the few bryophyte species that occur in the dryland zone, although is not confined to it. Associates include *Bryum argenteum, B. canariense, B. dichotomum, Weissia* spp. and *Chionoloma bombayensis*.

Status and distribution *T. brachydontium* is very widespread on the island except at high altitude. *Range*: Sub-cosmopolitan.

1 mm

LEFT **Moist shoots** RIGHT **Dry shoots**

Trichostomum crispulum

Pottiaceae

Synonym *Tortula crispula*

Description Plants yellowish-green to dark green, forming compact tufts or patches, shoots erect, 0.5–3 cm high. Leaves much curled when dry, spreading-erect when moist, 2–2.5 mm long, variable in shape, tapering to a hooded apex (shaped like the prow of a boat), the nerve ending in the apex; margins in upper part of leaf erect or somewhat incurved. Basal leaf cells rectangular, thin-walled (variable in extent), with a gradual transition to very small, opaque, highly papillose cells above. Fertile plants not reported from St Helena.

Recognition In typical forms, the hooded leaf apex is characteristic and is the best field character. This is not always pronounced, and it is safest to name only those collections in which it is well developed. Some forms of *Weissia* spp. can have a slightly hooded leaf apex, but the leaves are usually proportionately narrower and taper to a narrower, sharper point. When dry, the leaves of *T. crispulum* also tend to be more tightly curled than those of *Weissia* spp. The leaves of *T. brachydontium* are not hooded at the apex, and spread more widely when moist.

Habitats Confined to the drier parts of the island, growing on soil on hill-slopes and crags, sometimes by tracks and on road verges and walls, often in association with *T. brachydontium* and/or *Weissia* spp.

Status and distribution Probably frequent, and likely to have been widely overlooked because of the similarity of some of its forms to species of *Weissia*. *Range*: Widespread in Europe, local in Africa and Central and South America.

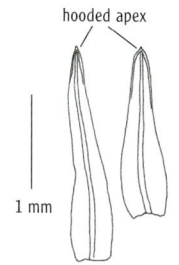

hooded apex

1 mm

LEFT **Moist shoots** TOP RIGHT **Sporophytes with long-beaked calyptrae** BOTTOM RIGHT **Dry shoots**

Weissia controversa

Pottiaceae

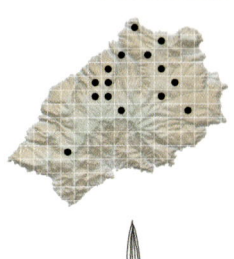

Description Plants forming small cushions or tufts; shoots 0.5–1 cm high, yellowish-green above, older parts darker. Leaves narrowly lanceolate, up to 3 mm long, margins rather strongly inrolled in the upper part of the leaf, nerve 30–60 μm wide at base, excurrent in a sharp point. Leaves strongly curled and crisped when dry, the upper part sometimes curled round like a crozier; when moist, leaves spreading-erect. Basal cells rectangular, grading to small, opaque, highly papillose cells above. Seta 6–10 mm long, yellowish; capsule ovoid-cylindrical, 0.5–0.7 mm long, mouth wide (in dehisced capsule), operculum long-rostrate; peristome present, but the teeth sometimes broken or short and difficult to see with a hand lens; spores 16–20 μm.

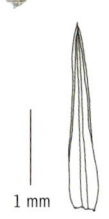

1 mm

Recognition Characteristic features include the yellowish-green colour, narrow leaves with strongly inrolled margins above, wide mouth of the capsule and the presence of a peristome. Only plants with mature capsules can be certainly identified, and key features should be checked under the microscope. *W. controversa* is very similar to the two other species described below. Forms of *Trichostomum crispulum* may resemble *Weissia* spp. (see account for distinctions).

Habitats *W. controversa* is a colonist of bare soil in a wide range of habitats, often growing with *Bryum argenteum* and *Trichostomum* spp. It is certainly more widespread than the map indicates, as suggested by the numerous and widespread records of sterile tufts.

Status and distribution *W. controversa* is frequent at low to mid altitudes on the island. *Range*: Sub-cosmopolitan in temperate climates.

Further species Two other species were recorded during the 2005 survey and, like *W. controversa*, cannot be identified in the absence of mature capsules. *W. brachycarpa* var. *obliqua* differs from *W. controversa* in the small capsule mouth covered by a membrane, lack of a peristome, larger spores (20–34 μm) and sometimes only weakly inrolled leaf margins. *W. condensa* also lacks a peristome, but the spores are small (14–20 μm) and the leaf nerve much broader (to 110 μm wide at base). Both species are found on open soil in a range of habitats. Both species have a sub-cosmopolitan range.

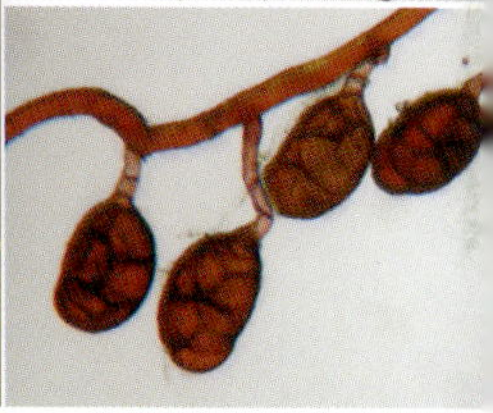

LEFT **Moist shoots** TOP RIGHT **Shoot showing rhizoidal tubers on subterranean rhizoids** BOTTOM RIGHT **Rhizoidal tubers**

Leptobryum pyriforme Meesiaceae

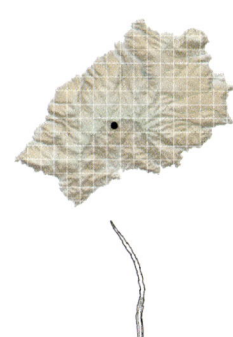
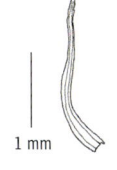

1 mm

Description Forming pale or yellowish-green, open patches or tufts of erect, slender plants to 2 cm high. Leaves very narrow, shrunken when dry, rapidly increasing in size up the stem, upper leaves to 3 mm long, composed mostly of nerve above, flexuose-spreading when moist, often crowded at the shoot apex; margins denticulate or more coarsely toothed above; nerve $1/3$ leaf width near base. Leaf cells narrowly rectangular throughout, 75–200 µm long × 10–12 µm wide, thin-walled. Brownish or purplish, ellipsoid tubers (100–150 µm long) are present, sometimes in abundance, on subterranean brown rhizoids, and are sometimes also present on short rhizoids on the lower parts of the stem. Capsules narrowly elongate-pear-shaped, inclined or pendulous, on a long seta, but not yet reported from the island.

Recognition *L. pyriforme* is a small species, with a rather weak, wispy appearance, and is rather easily recognised by the crowded tufts of narrow, long, flexuose, spreading, toothed leaves at the apex of the shoot, and the narrow cells throughout the leaf. The distinctive rhizoidal tubers provide a useful character, and if only subterranean, can be found by washing the soil from the plant.

Habitats Currently known only as a weed in flower pots in a garden and a glasshouse, growing on gritty or peaty compost. Elsewhere in its range, it is also known from a wide range of habitats including arable fields, paths, damp soil and rocks, and rotting wood.

Status and distribution Reported only from the gardens of Mt Pleasant. It is likely to be associated with cultivation elsewhere, and perhaps also occurs in semi-natural habitats on the island. *Range*: Sub-cosmopolitan, mainly in temperate regions.

LEFT Dry tuft with *Exormotheca pustulosa* in foreground RIGHT Leaves with long, hyaline apices

Bryum argenteum

Bryaceae

Description This small species forms 'pure' whitish or silvery mats or tufts on soil, or is sometimes found in mixed bryophyte mats. Shoots to 2 cm high (occasionally taller), cylindrical, whitish above, green to brownish below, sparingly branched. Leaves 0.8–1.2 mm long, strongly concave, the upper part hyaline, quickly or gradually narrowed to a short or long acute apex; margins flat, unbordered; nerve ending below apex or excurrent. Leaf cells rhomboid-hexagonal, 2–3 times as long as wide, marginal cells narrower. Fertile plants not reported from St Helena, but the capsule is small, pear-shaped or ellipsoid, inclined or pendulous, on a short seta.

Recognition This species is one of the most easily recognised on the island. The small, pale or whitish shoots and leaves that are contracted to a narrow apex (often looking like silvery hair-points when dry) are highly distinctive and unlike any other St Helena species. The leaves are fragile and easily split on the microscope slide.

Habitats Mainly on freely draining soils in a range of open to lightly-shaded habitats: on hill-slopes, on rocky ground and banks, by paths, on wall-tops, on broken tarmac and rocky soils on roadsides. Recorded associates include *Ceratodon purpureus*, *Bryum dichotomum*, *Trichostomum brachydontium* and *Campylopus* spp.

Status and distribution Widespread on the island, mainly at mid-altitudes. It is one of the few species that occur in the dryland zone and found, for example, on Rupert's Hill, in Broad Gut and near Prosperous Bay. *Range:* Cosmopolitan in temperate to cold environments.

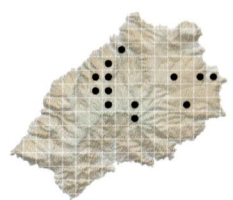

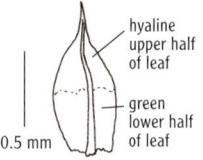

hyaline upper half of leaf

green lower half of leaf

0.5 mm

Colour forms of moist shoots

Bryum canariense

Bryaceae

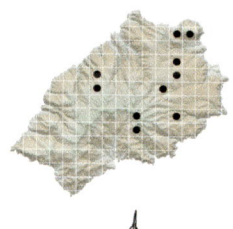

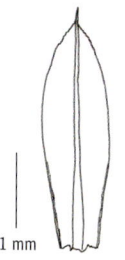

1 mm

Description Plants variable in size, colour and form: yellowish- to dark olive-green, sometimes dark reddish, forming tufts, cushions or mats, or found in mixed bryophyte mats. Shoots 1–4 cm high, stems reddish, frequently branched, with abundant red rhizoids below. Leaves 2.8–3.2 mm long, sometimes crowded at the apex of the shoot (a comal tuft), and on long shoots there may be two or more comal tufts at intervals along it; leaf margins narrowly recurved below, usually sharply toothed above; nerve sometimes reddish, excurrent in a reflexed point. Leaves when dry, glossy, somewhat appressed, shrunken, typically with the apical point strongly reflexed; when moist, spreading-erect, concave. Cells rectangular in leaf base, shortly hexagonal above, the marginal two or more rows narrow, but not forming a distinct border. Capsule pendulous, on a long red seta, but fertile plants not reported from the island.

Recognition The successive comal tufts, and the leaves appressed and very glossy when dry with a reflexed apical point are key field characters. Although variable in colour, *B. canariense* has a distinctive, robust appearance and is unlikely to be mistaken for any other on St Helena.

Habitats A plant of freely draining soils in open to moderately shaded places: by paths, on roadsides and waste ground, in open shrubberies and copses, on soil overlying rock and on a soil-capped wall. Recorded associates include *Bryum argenteum*, *Campylopus introflexus*, *Ceratodon purpureus*, *Trichostomum brachydontium* and *Weissia* spp.

Status and distribution Widespread, mostly in fairly dry habitats on St Helena, though not found in the driest parts of the dryland zone. *Range*: Western Europe and the Mediterranean area, Macaronesia, and locally in Africa and North and South America.

LEFT **Dry shoots with *Trichostomum brachydontium*** RIGHT **With abundant axillary bulbils**

Bryum dichotomum

Bryaceae

Synonym *Bryum bicolor*

Description Plants usually yellowish-green or pale green, reddish or darker below, forming small patches or scattered in mixed bryophyte mats; shoots often rather glossy when dry, 0.5–2 cm high, sparingly branched. Leaves appressed and somewhat crisped when dry, spreading-erect when moist, 1 mm long, very concave, acute; nerve excurrent in a usually long, green, yellowish, or often partly hyaline point up to 0.4 mm long; margins usually narrowly recurved below. Bulbils 1–3 per leaf axil, 250–350 μm long. Leaf cells square to rectangular below, narrowly hexagonal above, the marginal rows narrower but not forming a distinct border. Fertile plants not reported from the island: capsules pendulous, shortly ovoid, borne on a short seta.

Recognition Several species of *Bryum* develop bulbils in their leaf axils, but since *B. dichotomum* is the only one known from St Helena, the presence of axillary bulbils (easily seen with a hand lens) immediately identifies it. *B. dichotomum* is a variable species, with forms having blunt or acute leaves, a nerve excurrent or not, and a variable number of axillary bulbils. However, it appears to be fairly uniform on St Helena, and the leaves almost always have a long-excurrent nerve which, when partly hyaline, can give the dry plant a rather hoary appearance.

Habitats In tufts or scattered amongst other bryophytes, on freely draining soil in open or lightly shaded habitats: on earthy and rocky roadside banks, on paths and rocky ground, in cultivated fields and flower beds. Also found on walls and soil overlying rock outcrops.

Status and distribution Widespread at middle altitudes on St Helena. *Range*: Widely distributed in the temperate regions of Europe, Asia, Macaronesia, Africa and North America.

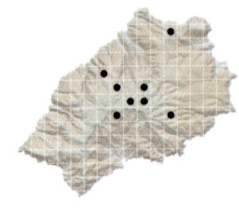

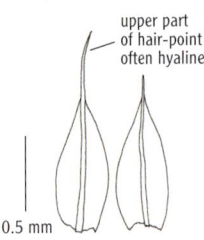

upper part of hair-point often hyaline

0.5 mm

LEFT **Moist shoots** TOP RIGHT **Shoot showing tubers on subterranean rhizoids** BOTTOM RIGHT **Rhizoidal tubers**

Bryum klinggraeffii

Bryaceae

Description A small, anonymous-looking *Bryum*, growing in pure or mixed, loose or compact tufts, one of a small group of related species in the *B. erythrocarpum* group. Shoots 0.5–1.0 cm high. Leaves shrunken when dry, spreading-erect when moist, 1–1.5 mm long; margins narrowly recurved below, denticulate above. Leaf cells rectangular near the leaf base, otherwise narrowly hexagonal, the marginal cells narrower but not forming a distinct border. Rhizoids pale yellowish to pale brown, with abundant small, red, irregularly rounded tubers, 70–105 µm long. Fertile plants not observed on St Helena.

Recognition Except sometimes for *B. rubens*, members of the *B. erythrocarpum* group cannot be named in the field, but only by observing the rhizoidal tubers with a high power microscope (carefully wash away soil from the plants). The small, red, irregular tubers on pale rhizoids will immediately identify *B. klinggraeffii*. *B. rubens* and *B. subapiculatum* have much larger tubers (see accounts, p. 100 and p. 102).

Habitats *B. klinggraeffii* has been recorded only as a weed in pots of young gumwood plants, with *Bryum rubens*, at the Conservation Nursery, Scotland, and may be a recent introduction. Elsewhere in its range, it is found in arable fields, on open banks and on the muddy margins of ponds.

Status and distribution At present, known from only one location. It is possible that *B. klinggraeffii* and related species from pots might be found transplanted with gumwood and other endemic species in newly established plantations. *Range*: Europe, Asia, North America and Patagonia.

Further species A very small amount of a related species, *B. radiculosum*, has been recorded from the mortar of a stone wall near Plantation House. It is identified by the yellowish to brown, coarsely papillose rhizoids with larger, spherical, red or brown tubers, 120–180 µm long. The strong leaf nerve is also characteristic.

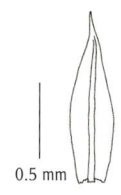

0.5 mm

LEFT **Moist shoots** TOP RIGHT **Shoot showing rhizoidal tubers on aerial and subterranean part of stem** BOTTOM RIGHT **Rhizoidal tubers**

Bryum rubens

Bryaceae

Description A member of the *B. erythrocarpum* group, forming tufts 0.5–1.5 cm high, or diffuse patches; lower parts of the shoots reddish. Leaves shrivelled and somewhat crisped when dry, spreading-erect when moist, 1–2 mm long, margins narrowly recurved below, flat above, and becoming sharply denticulate near the apex. Basal leaf cells narrowly rectangular, then narrowly hexagonal above; marginal cells longer and narrower with thicker walls, and forming a distinct border. Tubers spherical, crimson or red (white when young), to 250 μm or more, with strongly convex cells, developed on very short rhizoids in the lower part of the shoot and around its base, and on longer subterranean rhizoids (sometimes only on the latter). Rhizoids brown, papillose. Fertile plants not observed on St Helena.

Recognition Plants of *B. rubens* that develop tubers clustered around the base of the stem can be identified in the field with a ×20 hand-lens. However, tubers may not always be obvious or be present only on subterranean rhizoids, so that some collections will need to be confirmed by microscopic examination. Furthermore, two or more tuberous *Bryum* species may grow in close proximity with their rhizoids intertwined, and *B. subapiculatum* may occasionally develop aerial tubers (for differences, see account of the latter). In all other members of the group, the tubers occur only on long, subterranean rhizoids.

Habitats On freely draining soil in pots of cultivated plants at the Conservation Nursery, Scotland, and on light soil in a flower bed at Mt Pleasant.

Status and distribution Currently recorded from only two locations, but doubtless occurs elsewhere in plant pots and in cultivated ground. *Range*: Europe, also reported from Macaronesia, South Africa, North America and Australasia.

0.5 mm

LEFT **Moist shoots** TOP RIGHT **Shoot showing tubers on subterranean rhizoids** BOTTOM RIGHT **Rhizoidal tubers**

Bryum sauteri

Bryaceae

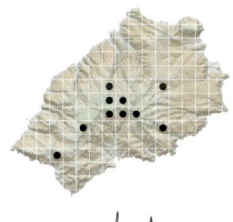

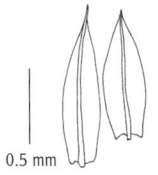

Description *B. sauteri*, a member of the *B. erythrocarpum* group, forms tufts or patches 0.5–1.5 cm high, mostly on bare soil, the tufts often more compact than those of other members of the group. Leaves 0.8–1.2 mm long, shrunken when dry, spreading-erect when moist; margins narrowly recurved below; nerve rather strong, usually excurrent in a short point. Leaf cells narrowly rectangular below, hexagonal above, the marginal cells narrower but not forming a border. Rhizoids reddish-brown, finely papillose, bearing small (60–100 × 40–60 µm), pear-shaped, brown or reddish tubers on long rhizoids. Fertile plants have not been observed on St Helena.

Recognition This species resembles other tuberous *Bryum* species vegetatively, and cannot be named in the field. However, the small, reddish-brown, pear-shaped rhizoidal tubers, composed of few cells, are diagnostic. Care is needed in washing out the soil from the plants as the tubers are small, can be rather easily detached from the rhizoids, and may be few in number.

Habitats A plant of freely-draining or moisture-retentive soils in open or partly shaded habitats: on roadside banks and verges, by paths, on woodland banks, at the base of rock outcrops, in garden flower pots and once recorded growing in a crevice on the trunk of a mature Caffra thorn tree, an exceptional habitat.

Status and distribution The most frequent and widespread of the tuberous *Bryum* species on St Helena, from mid-altitudes ascending to the Depot and Diana's Peak National Park. *Range*: Europe, Macaronesia, South Africa, South America and Australasia.

0.5 mm

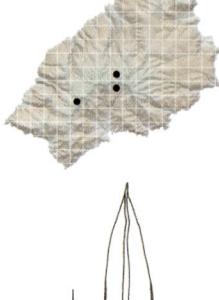

LEFT Moist shoots TOP RIGHT Shoot showing tubers on subterranean rhizoids BOTTOM RIGHT Rhizoidal tuber

Bryum subapiculatum   **Bryaceae**

Synonym *Bryum microerythrocarpum*

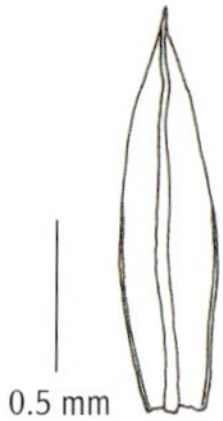

Description *B. subapiculatum* is another of the small, rather featureless mosses
that comprise the *B. erythrocarpum* group. It grows in tufts 0.5–1.5 cm high,
or in small, loose patches on bare soil. Leaves shrunken when dry, spreading-
erect when moist, 1–1.5 mm long, narrowly lanceolate to elongate-ovate,
acute; margins narrowly recurved below, flat and denticulate above; nerve
rather strong, usually shortly excurrent. Leaf cells rectangular at base,
narrowly hexagonal above, the marginal cells narrower but not forming a
distinct border. Rhizoids brownish, papillose. Tubers on long rhizoids
(occasionally aerial), large (180–260 µm), rounded, scarlet or brick-red, the
cells not or only slightly convex. Fertile plants not observed on St Helena.

Recognition *B. subapiculatum* is a variable species, but characterised by the flat
(or at most, weakly convex) cells of the large tubers, and by the leaves lacking
a distinct border of narrow cells. *B. rubens* differs in the convex-bulging cells
of the tubers, and the bordered leaves. In *B. rubens*, the tubers are
predominantly aerial or clustered around the shoot base, but they are mostly
or entirely on long rhizoids in *B. subapiculatum*, although sometimes very few
in number.

0.5 mm

Habitats On bare soil on a grassy bank, at the base of a low roadside wall and on a soil bank by a stand of New
Zealand flax.

Status and distribution There are only three records from St Helena, all at rather high altitudes: near High Peak,
by the Sandy Bay Ridge road and in Diana's Peak National Park. However, it must surely occur also at lower
altitudes on the island. *Range*: Europe, western Asia, North America and Australasia.

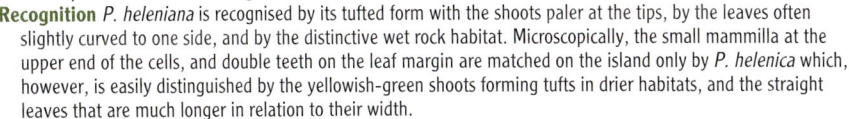

LEFT & CENTRE **In situ, Wash House Gut** RIGHT **Variation in leaf form**

Philonotis heleniana

Bartramiaceae

Synonym *Bartramia heleniana*

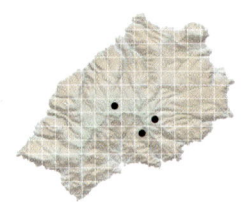

Description Plants in small tufts attached to damp rocks, green with paler tips, densely matted with brown tomentum below. Shoots ascending to pendent, the apices often curved or hooked, short forms to 2–3 cm long and sparsely branched, tall forms to 6 cm long with frequent short branches near shoot apex. Leaves crowded, triangular-lanceolate, 1 mm long, weakly to strongly curved to one side, nerve strong and toothed on the back, excurrent in a sharp apex, leaf margin with double teeth for most of its length. Leaf cells rectangular: on the upper surface of the leaf, each cell with a small, bulging prominence (mammilla) at its upper end. Capsules of *P. heleniana* have not been described, but capsules of other species of *Philonotis* are globose and borne on a long seta.

Recognition *P. heleniana* is recognised by its tufted form with the shoots paler at the tips, by the leaves often slightly curved to one side, and by the distinctive wet rock habitat. Microscopically, the small mammilla at the upper end of the cells, and double teeth on the leaf margin are matched on the island only by *P. helenica* which, however, is easily distinguished by the yellowish-green shoots forming tufts in drier habitats, and the straight leaves that are much longer in relation to their width.

Habitats On wet rocks in and by streams (Wash House Gut and upper Deep Valley) and near a waterfall below the Sandy Bay Ridge road.

Status and distribution Apparently endemic. Although currently known from only three locations, it is likely that *P. heleniana* will be found on wet rocks in other sites. Many species of *Philonotis* are extremely variable, and this notoriously difficult genus requires taxonomic revision in many parts of the world, including in Africa. It is possible that a revision will show that *P. heleniana* is synonymous with species known from elsewhere.

LEFT **In situ on roadside bank** RIGHT **Leaf form**

Philonotis helenica

Bartramiaceae

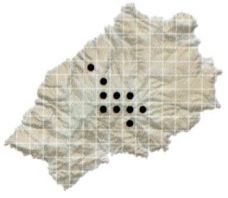

Description Plants forming pale or yellowish-green tufts or mats, with foliage brownish below; shoots 0.5–3 cm high, the stems with matted tomentum, branches few. Leaves narrowly triangular, straight or slightly curved to one side, crowded on the stem, 1–1.4 mm long, the nerve excurrent in a long, sharp apex, margins with sharp, double teeth from base to apex. Leaf cells elongate-rectangular, each with a small mammilla (bulging prominence) at the upper end of the cell, though the mammillae may be obscure or apparently absent in young leaves. Capsule globose, on a seta 8–10 mm long. Weak plants may have smaller and more widely spaced leaves, with a weak nerve disappearing below the leaf apex.

Recognition The pale or yellowish colour of the shoots, the tufted growth, the long narrow leaves, and the habitat together characterise this species, which can usually be easily recognised in the field. It can, however, also occur as scattered shoots in mixed bryophyte communities. Microscopically, the toothed leaf margin and small mammilla in each cell are key characters of the genus. The distinctive globose capsules are present in a collection made by Balansa in 1877, but capsules have not been reported since.

Habitats Typical habitats include earthy banks, especially the steep, shaded soil exposures on roadside banks and walls, growing in open locations or hidden beneath hanging vegetation. It is also found on soil covering rocks in shaded locations.

Status and distribution Apparently endemic. This species is widespread on the island at middle and upper altitudes. *P. helenica* closely resembles other described species from South America and Africa, and it is possible that a revision of this difficult genus will show that *P. helenica* is synonymous with species described from elsewhere.

TOP LEFT & RIGHT **Dry shoots** BOTTOM LEFT & RIGHT **Moist shoots with sporophytes, showing the typical cladocarpous form (right)**

Macrocoma tenuis subsp. *tenuis*

Orthotrichaceae

Synonym *Macromitrium microphyllum*

Description Plants olive, dull green to brownish below (young shoots yellowish-green), forming extensive, dense mats on the trunks of trees. Stems creeping, pinnately branched, the abundant branches spreading to ascending, some short, but many branching again to form secondary pinnate stems. Leaves 1–1.3 mm long, ovate-lanceolate, margins narrowly recurved below; nerve strong, disappearing below apex; when dry, leaves tightly appressed-erect and only slightly curved; when moist, spreading, channelled, with incurved apices. Leaf cells small, rounded, strongly bulging-papillose. Fertile plants common: seta yellowish, 4–5 mm long, capsules erect, shortly cylindrical, 1 mm long, furrowed when dry, with a long neck; calyptra long-conical, covering the whole capsule.

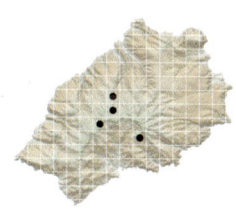

Recognition This species is easily recognised by the creeping, pinnate shoots, with abundant spreading or ascending branches, and small leaves that are closely appressed when dry. Species of *Macromitrium* also have creeping stems with abundant, crowded short branches, but are more robust plants with longer leaves that are crisped when dry. *M. urceolatum* further differs in its fragile leaf apices.

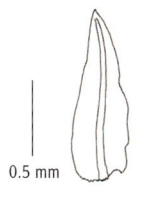

0.5 mm

Habitats Mainly an epiphyte on the trunks of trees, forming dense, flat mats of shoots, and often with abundant capsules. It is often associated with *Macromitrium* spp. There is one record from a dry rock face (Wash House Gut).

Status and distribution At middle altitudes on the island: reported from Fairyland, Wash House Gut, Cason's, Plantation House and near Bishop's Bridge. *Range*: *M. tenuis* subsp. *tenuis* is widespread and locally frequent in Africa, and also occurs in Australia and New Zealand; subsp. *sullivantii* occurs in west and east Asia, and South America.

LEFT & RIGHT **Moist shoots**

Daltonia splachnoides

Daltoniaceae

Description Plants 4–7 mm high, in small tufts or as isolated shoots. Stems dark brown. Leaves mid- to yellowish-green, to 1 mm long, keeled, uniform in size in upper half of shoot; glossy when dry, spreading at 45° from stem when moist, somewhat appressed and crisped when dry; nerve disappearing towards leaf apex; margins flat. Laminal cells 20–40 μm long × 6–12 μm wide, variable in shape; border strong, composed of very long, narrow, thick-walled cells (150 μm long × 6 μm wide), widening from 2–3 rows in upper part of leaf to 8–10 rows near leaf base; inner cells elongate-hexagonal. Fertile plants not yet found on the island. However, the calyptra is highly distinctive in its wide basal fringe that is deeply divided into many narrow segments.

1 mm

Recognition Although the species is small and difficult to find (especially isolated shoots), the combination of erect shoots, long, narrow, sharply pointed leaves with short laminal cells and a stout border of linear cells render this species unmistakable.

Habitats On the trunk of a 10-year old planted whitewood, associated with *Cheilolejeunea ascensionis* and *Dendroceros adglutinatus*, and on a fallen branch of a Norfolk Island pine, with *Dendroceros adglutinatus* and *Lejeunea autoica*. Elsewhere in its range, known also from wet rock and humus.

Status and distribution Apparently very rare; found on only two trees in Diana's Peak National Park, despite careful searches of tree bark in many locations there. The presence of several tufts on a fallen branch of the Norfolk Island pine suggests that its main habitat on the island might be the canopy branches, and that this tree may be the source of the spores that germinated on the nursery-grown whitewood after it had been planted out. *Range*: Western Europe, Macaronesia, North and Central America, China and New Zealand.

LEFT **Moist shoots (some perianths of *Chiloscyphus coadunatus* showing)** RIGHT **Dry shoots**

Lepidopilidium crispifolium

Pilotrichaceae

Description Plants mid- to deep dull green, the youngest shoots paler; shoots prostrate, rather flattened, densely and sub-pinnately branched, the branches short. Leaves crowded, slightly glossy, strongly shrunken and undulate-crisped (contorted) when dry, the nerve double, diverging, usually indistinct, disappearing below mid-leaf. Leaves of two kinds: those on the upper side of the shoot (dorsal leaves) rather broad, 1.2–1.4 mm long, somewhat concave, elevated from the stem when moist; those on lower side of shoot (lateral leaves) longer and more gradually tapering, the lower part of the leaf incurved and forming a deep fold on one side. Leaf cells 30–60 μm long × 8–11 μm wide in mid-leaf (4–6 times as long as wide). In fertile plants, seta reddish, 10 mm long; capsule ovoid, 1.5 mm long, inclined, operculum sharply rostrate.

Recognition This species is most readily differentiated from *L. pallidifolium* by the notably shrunken, crisped and crinkled appearance of the leaves in the dry shoot, and in the moist shoot by the broader, more abruptly pointed dorsal leaves which are often raised up from the stem giving the plant a rather spiky appearance. Shorter leaf cells also differentiate *L. crispifolium* from *L. pallidifolium*.

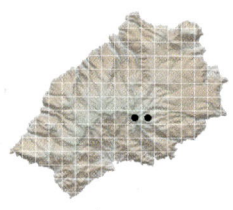

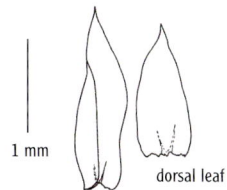

1 mm

dorsal leaf

lateral leaf

Habitats A plant of deeply shaded, moist environments, growing on rocks and soil near seepages, under ferns or shrubs, and on wet rocks by streams in deep guts and defiles. It may grow in small, 'pure' mats of interwoven shoots, or in mixed bryophyte communities.

Status and distribution Endemic. *L. crispifolium* is currently known from only a few sites on the southern and western slopes of Diana's Peak National Park. It is presumably to be found elsewhere, but might be confined to the central Peaks area.

LEFT **Moist shoots** RIGHT **Dry shoots**

Lepidopilidium pallidifolium

Pilotrichaceae

Synonyms *Hookeria pallidifolia, Hookeriopsis pallidifolia, Lepidopilidium virens*

Description Plants variable, pale to deep green or yellowish-green, glossy, soft, forming dense mats: shoots flattened, creeping, variable, to 4 mm wide, with sparse and irregular long branches, or sub-pinnate with many short branches. Leaves asymmetrical, those on the upper side of the shoot (dorsal leaves) and the lateral leaves somewhat differing in shape, but the differences are not as marked as in *L. crispifolium*; the lower part of the lateral leaves is incurved and forms a fold on one side; leaves to 2 mm long, margins weakly toothed towards apex; nerve short and double, but usually very obscure or apparently absent. Cells in mid-leaf long and narrow, 110–150 µm long × 8–12 µm wide (7–10 times as long as wide), in leaf apex much shorter, elongate-hexagonal. Fertile plants frequent: seta red, 7–11 mm long; capsule inclined, ovoid, the operculum sharply rostrate.

Recognition The flattened pale or yellowish-green shoots, not strongly undulate-crisped when dry, and leaves often curved downwards, are key characters for field recognition. It is a variable species, with many colour and morphological forms. Microscopically, the mid-leaf cells are much longer than those of *L. crispifolium*. The unrelated *Isopterygium* sp. also has pale, flattened shoots, but is much smaller with extremely narrow leaf cells.

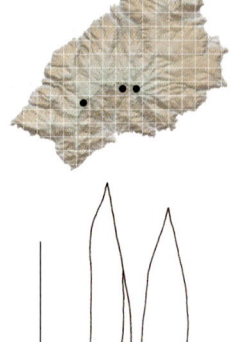

1 mm

lateral leaf   dorsal leaf

Habitats In moist, shaded locations, on peaty soil, soil overlying steep rock cliffs and banks, on low rocks and in rock crevices, on rotting tree trunks, branches and other vegetation.

Status and distribution *L. pallidifolium* appears to be restricted to the Central Ridge, with most records from Diana's Peak National Park and others from the Depot. It is perhaps to be found also at High Peak and in other shaded localities at high altitudes. *Range*: Otherwise known only from the Azores, Madeira and southernmost Spain (formerly as *Lepidopilidium virens*).

Moist shoots TOP LEFT **Deep wefts on sandy bank** TOP RIGHT **Pinnate branching** BOTTOM RIGHT **Forms of stem and branch leaves**

Kindbergia praelonga

Brachytheciaceae

Synonyms *Hypnum praelongum, Eurhynchium praelongum*

Description Plants yellowish-green to dull, dark green, slender, creeping, elongate, rather regularly pinnate and bipinnate, branches short or elongate, usually flattened in one plane. Leaves shrunken when dry, spreading when moist, acute, margins strongly denticulate to leaf base, nerve $^1/_2$–$^3/_4$ leaf length. Leaves of two kinds: stem leaves to 1.8 mm long, broad, quickly tapering above to an often long, fine apex, widely clasping the stem and decurrent at base; branch leaves lanceolate, acute, about 1 mm long. Leaf cells narrow, 5–12 times as long as wide, but towards base wider and shorter. Fertile plants not yet reported on St Helena: seta long, capsule ovoid, the operculum long-rostrate.

Recognition Although variable, *K. praelonga* is easily recognised by the slender, elongate, straggling wefts of usually flattened, bipinnate shoots, and the broad stem leaves that strongly differ in shape from the narrower branch leaves. In the related *Oxyrrhynchium hians*, the stems are not regularly pinnate, the stem and branch leaves differ only slightly in shape and they taper to a shorter apex.

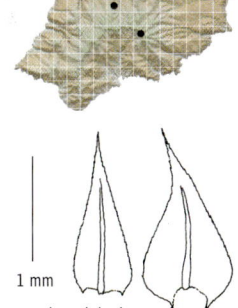

1 mm

branch leaf

stem leaf

Habitats On soil in shaded or open locations, on roadsides, path-sides and woodland banks, often in large patches. Elsewhere in its range, also on tree trunks, logs, by waterways and cliff ledges.

Status and distribution Apparently local on St Helena, with records only from Plantation House woods and roadside banks nearby, and by a grassy path in Diana's Peak National Park. *Range*: Very widespread in the temperate northern hemisphere, extending to north Africa. Also known from South Africa and Australasia, perhaps as an introduction. Like *Pseudoscleropodium purum*, it may be also be an introduction to St Helena, perhaps as a packing for imported plants.

LEFT Moist shoots RIGHT Irregularly pinnate branching

Oxyrrhynchium hians

Brachytheciaceae

Synonym *Eurhynchium hians*

Description Plants usually yellowish-green, dull to slightly glossy, creeping, elongate, irregularly pinnate or bipinnate, the branches elongate. Leaves 1–1.5 mm long, moderately shrunken when dry, spreading when moist, acute, margins dentate from leaf base to apex, nerve $^1/_2$–$^3/_4$ leaf length, stem and branch leaves ovate, similar in size and shape (the stem leaves slightly wider and narrowly clasping the stem), tapering to an acute apex. Leaf cells narrow, 4–10 times as long as wide, towards the base wider and shorter. Fertile plants not reported: seta long, capsule ovoid, the operculum shortly rostrate.

Recognition *O. hians* may be recognised by the usually yellowish-green colour, the often rather 'springy' wefts with branches not flattened in one plane, and the widely ovate stem and branch leaves that are similar in shape. It is unlikely that this species would be mistaken for *Kindbergia praelonga* if attention is paid to the respective shapes of stem and branch leaves. However, it is possible that other species of this or related genera could occur on St Helena.

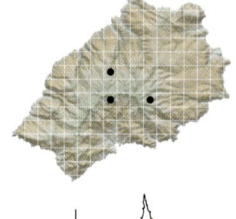

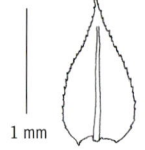

1 mm

Habitats On soil on a woodland bank, and grassy vehicle tracks on a driveway. Elsewhere in its range, it is found in pastures, by streams, and on rocks and walls.

Status and distribution Apparently local on St Helena, but presumably under-recorded: the only records are from near the Boer Cemetery, Plantation Wood and Mount Pleasant. *Range*: Europe to central Asia, Macaronesia, north, west and east Africa, and North America.

TOP LEFT **Dry shoots** BOTTOM LEFT & RIGHT **Moist shoots with characteristic swollen appearance (right)**

Pseudoscleropodium purum

Brachytheciaceae

Synonym *Hypnum purum, Scleropodium purum*

Description Plants large, forming coarse wefts, carpets or mounds that are sometimes large and extensive, usually pale green or yellowish-green, sometimes brownish; shoots cylindrical (swollen-looking), to 15 cm long, pinnately branched, procumbent or ascending. Leaves 2 mm long, closely overlapping, deeply concave, with a very small, sharp, abrupt, often reflexed point at the rounded apex, margins flat or narrowly recurved at base, finely denticulate above. Leaf cells long and narrow, except at the base. Fertile plants not reported from St Helena.

Recognition This highly distinctive species is easily recognised by the swollen shoots, by the broad, concave leaves with a short apical point and by its carpet or mound-forming habit. With the hand lens, the sharp leaf point is most easily seen at the apex of the shoot.

Habitats An invasive species, forming dense carpets or mounds in open or partially shaded places, on banks, along grassy paths, on roadside verges and locally on the ground in open woodland. In Diana's Peak National Park, locally spreading from the paths into native habitats, overgrowing and eliminating native species.

Status and distribution Locally abundant in some areas of Diana's Peak National Park, particularly along Cabbage Tree Road and the Ridge Path. Common elsewhere above 500 m. The encroachment of *P. purum* from its extensive stands along the grassy path in Diana's Peak National Park into native cloud forest communities indicates that physical removal should be undertaken as a conservation measure. *P. purum* is known to have been used as packing material for imported plants, and almost certainly should be treated as a non-native introduction on St Helena. *Range*: Considered native to Europe and Macaronesia, but widely known elsewhere as a probable introduction, including in east Asia, Sri Lanka, Australasia, and North and South America.

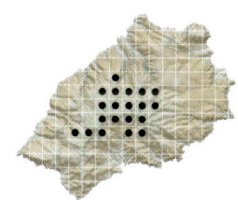

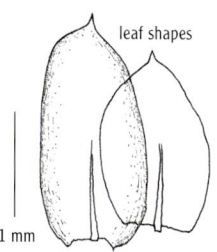

leaf shapes

1 mm

Moist shoots LEFT **In situ** TOP RIGHT Crowded branching BOTTOM RIGHT Detached shoot with mature sporophyte

Sainthelenia athroclada

Brachytheciaceae

Synonyms *Hypnum athrocladum, Brachythecium athrocladum*

Description Plants small, forming compact mats, the main stems creeping, with abundant, short, closely-packed, spreading or ascending branches. Leaves straight to weakly curved, appressed when dry, spreading when moist. Leaves of two kinds: stem leaves triangular below, lanceolate above, 0.7–1.1 mm long, decurrent, the base minutely toothed with projecting cell walls; branch leaves narrower and smaller, 0.5–0.8 mm long. Cells in mid-leaf 25–55 μm long × 5–7 μm wide, thick-walled, and at the basal angles, nearly square or shortly rectangular. Fertile plants: seta 5 mm long, strongly papillose; capsule 1–1.3 mm long, suberect to horizontal, the operculum shortly rostrate.

Recognition This species can be recognised in the field by the small, prostrate plants forming dense mats, the crowded branches and small leaves, and when fertile by the short, papillose seta (appears rough when viewed with hand lens).

Habitats Grows on trees and rocks. At High Peak, found on a moist cliff ledge, at the Depot on the bark of a Mexican cypress with *Lejeunea autoica* and *Radula fulvifolia*, and at Wash House Gut, on rather dry rocks. Near Hutt's Gate, it grew on the trunk and base of a large roadside Caffra thorn tree, and below Cason's on the lower trunk of a large Monterey cypress.

Status and distribution Endemic. It is currently known only from the five locations mentioned above. First collected by W. Burchell between 1805 and 1810 from an unknown location, and later described by Mitten, as *Hypnum athrocladum*.

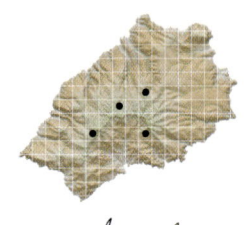

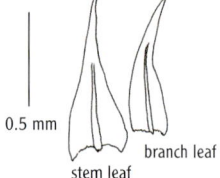

0.5 mm

stem leaf · branch leaf

Moist shoots in dense weft

Hypnum cupressiforme

Hypnaceae

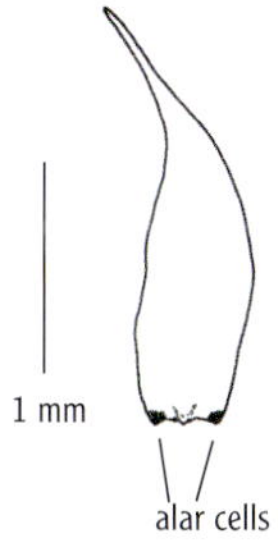

Description Plants medium-sized, usually mid-green, glossy when dry, procumbent, mat-forming, irregularly but freely branched. Rhizoids pale brown, in clusters along creeping stems. Leaves ovate-lanceolate, little altered when dry, concave, strongly curved down towards the underside of the shoot, curved to one side, tapering to a long, fine apex; margins entire; nerve very short and double, obscure or apparently lacking; stem leaves 1.7–1.9 mm long × 0.7 mm wide; branch leaves shorter and narrower. In most of the leaf, laminal cells linear, very narrow, but alar cells (at each side of leaf base) strongly differentiated, nearly square, colourless to pale brown (Fig. 8.21, p. 123). Capsule cylindrical, curved, on a long seta, the operculum rostrate, but fertile plants have not been observed on St Helena.

Recognition *H. cupressiforme* is characterised by its freely branched shoots with crowded, strongly concave, finely pointed leaves turned to one side and lacking a nerve (or having only an obscure forked one). It can be separated in the field from the closely related *H. lacunosum* by its colour, smaller size and more flattened branches, and from *H. jutlandicum* by its irregular branching.

Habitats On the top of a small rock on a steep roadside bank near Bishop's Bridge, associated with *Leptophascum leptophyllum*, *Fissidens pygmaeus* and *Cephaloziella* sp. Also on stone steps and on soil in Plantation Wood. Elsewhere in its range, it is found on tree trunks, logs, rotting wood, walls and rocks, rarely on soil.

Status and distribution Apparently very local on St Helena, and currently reported from only three sites, all within a single 1 km grid square. *Range*: A locally abundant sub-cosmopolitan species.

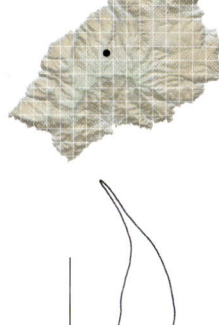

1 mm

alar cells

Moist shoots in dense weft

Hypnum jutlandicum

Hypnaceae

Synonym *Hypnum ericetorum*

Description Plants medium-sized, pale green, glossy when dry, forming flat mats, freely branched, the branching rather regularly pinnate. Rhizoids pale brown, in clusters along the creeping stems. Leaves ovate-lanceolate, little altered when dry, concave, strongly curved towards the underside of the shoot (often nearly in a semi-circle), curved to one side, narrowed above (sometimes rather abruptly) to a short or fairly long, narrow apex; margins obscurely denticulate; nerve very short and double, or obscure or lacking; stem leaves 1.6–1.8 mm long, branch leaves similar but smaller. Alar cells nearly square, colourless, strongly differentiated from the narrow, linear cells in the rest of the leaf. Fertile plants not observed on St Helena: seta long; capsules cylindrical, curved, the operculum rostrate.

Recognition The flattened, pale green shoots, the mostly nerveless, down-curved leaves turned to one side, and the rather regularly pinnate shoots identify this species in the field. *H. cupressiforme* has less regular branching and is mid-green in colour, and *H. lacunosum* is more robust, with cylindrical shoots and with stronger coloration.

Habitats On St Helena, only on peaty soil in an open area in woodland. Elsewhere in its range, on soil in upland grasslands, heathland and open woodland, on logs and rock.

Status and distribution Very local on St Helena, and currently known only from Plantation Wood. *Range*: Temperate to cold environments in Europe eastwards to western Asia, Macaronesia, east and west Africa, and North America.

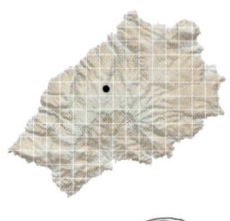

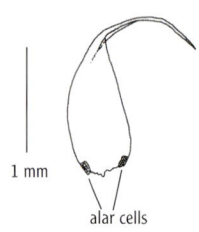

1 mm

alar cells

Moist shoots

Hypnum lacunosum

Hypnaceae

Description Plants rather robust, strongly coloured, usually golden brownish-green, glossy when dry, procumbent in mats, freely and irregularly branched, the branches cylindrical (looking swollen when moist), with closely overlapping leaves. Rhizoids pale brown, in clusters along the creeping stems. Leaves ovate or ovate-lanceolate, little altered when dry, concave, curved downwards, tapering from near base to a long, fine apex; margins obscurely denticulate; nerve very short and double, obscure or lacking; branch leaves to 2.4 mm long × 0.7 mm wide; stem leaves similar but often less curved to one side. In most of the leaf, laminal cells linear, very narrow (cell cavity only 2–3 µm wide), the alar cells strongly differentiated, rounded-square, colourless or pale brown. Fertile plants not observed on St Helena; seta long; capsules cylindrical, curved, the operculum rostrate.

Recognition *H. lacunosum* can be recognised by its rather large size, golden brownish-green colour, swollen shoots and leaves curved to one side. The St Helena taxon is not as robust as some forms of the species, and could perhaps be assigned to var. *tectorum*. The differential characters of the related species *H. cupressiforme* and *H. jutlandicum* are given in their accounts.

Habitats On St Helena, the only current report is from freely draining soil on open banks by a dirt road, with *Campylopus introflexus* and *Bryum canariense*. Elsewhere in its range, also found on rocks and sand dunes.

Status and distribution Apparently very local on St Helena, recorded only from the track between Rock Cottage and the sawmill. *Range*: Temperate and cool environments in Europe eastwards to western Asia, Macaronesia, eastern and southern Africa, and North America.

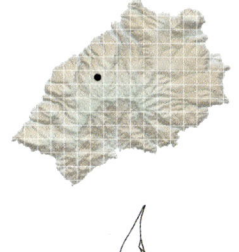

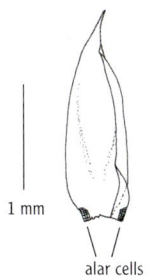

1 mm

alar cells

Moist shoot

Entodon dregeanus

Entodontaceae

Description Plants procumbent, rather rigid, branched, yellowish-green above, olive-green or brownish below. Stems yellowish-brown, strong, irregularly pinnate, the branches sometimes curved downwards. Leaves glossy, concave, moderately appressed but not shrunken when dry, slightly more spreading when moist, margins flat, minutely denticulate; nerve very short and double, or almost absent. Stem leaves symmetrical, 1.8 mm long; branch leaves 1.5–1.8 mm long, rather narrower, often slightly asymmetrical with the upper part slightly turned to one side. Most of leaf with linear cells, 90–120 μm long × 6 μm wide, but the well differentiated alar region is composed of square or shortly rectangular cells. Fertile plants not observed.

Recognition *E. dregeanus* is characterised by the rather robust, procumbent, much-branched shoots, glossy leaves, ovate-lanceolate symmetrical stem leaves with a nerve very short or lacking, and very long, narrow leaf cells. Of the other procumbent species in which the leaf nerve is very short or lacking, species of *Hypnum* differ in their leaves rather strongly turned to one side and tapering to a fine point, and *Sematophyllum* and *Isopterygium* are slender plants.

Habitats On a rotting log in an open area of Mexican cypress woodland at the Depot.

Status and distribution *E. dregeanus* is currently known from only one site, and should be looked for elsewhere in suitable habitats along the Central Ridge. *Range*: Widespread in west and east Africa, Madagascar and the Mascarene islands. It should be noted that some *Entodon* species are vegetatively similar, and are best separated by their peristome features. Unfortunately, capsules of the St Helena plant have not been found, but should be searched for so that its identity can be confirmed.

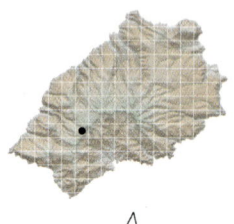

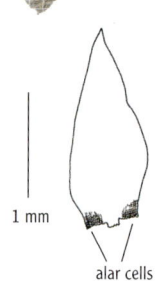

1 mm

alar cells

LEFT Moist shoots in dense weft RIGHT Close-up of shoot showing asymmetric leaves

Isopterygium sp.

Pylasiadelphaceae

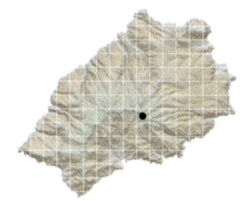

Description Plants pale, whitish- to yellowish-green, flattened, procumbent, forming loose, thin, soft mats, or in mixed bryophyte patches. Shoots slender, to 2 mm wide, simple or sparingly branched, often tapering and becoming very slender above with very small, narrow leaves. Rhizoids smooth, crimson, tending to be clustered near the leaf insertions. Leaves flat, glossy, shrunken when dry, strongly asymmetric, to 1.2 mm long; margins flat, minutely serrate above; nerve forked, very short. Leaf cells linear, very long and narrow throughout, 125–180 μm long × 5 μm wide (25–35 times as long as wide), a very few cells at leaf base shorter but hardly forming a distinct alar group. Fertile plants not observed.

Recognition The narrow, pale-coloured, flattened, soft shoots, and asymmetric, acute leaves are characteristic, but the identification of this plant should be confirmed microscopically by observing the extremely narrow and long leaf cells. *Lepidopilidium pallidifolium* is a much more robust plant, with longer, glossy leaves, and wider leaf cells that are long-hexagonal in the leaf apex. However, weak, isolated shoots might superficially resemble *Isopterygium*.

Habitats In small patches, or as scattered shoots creeping amongst or over other bryophytes, on humus-rich soil on rock, in crevices, or under overhangs, always in deeply shaded, humid locations in the cloud forest communities. Recorded associates include *Fissidens chioneurus* and *Kurzia nemoides*.

Status and distribution This plant has been recorded only from Diana's Peak National Park, in locations between Mt Actaeon and Cuckold's Point. However, it cannot be named at the present time. Latin American species of *Isopterygium* have been revised recently from 92 names down to 8 accepted species, but the genus remains very poorly known in Africa, where most of the 64 described species would probably be placed in synonymy when revised, leaving just a few 'good' species.

LEFT **Typical hanging weft** TOP RIGHT **Moist shoot** BOTTOM RIGHT **Sporophytes**

Sematophyllum erythrocaulon

Sematophyllaceae

Synonyms *Hypnum erythrocaulon, Sematophyllum plumarium*

Description Plants forming compact or loose mats, usually pale or yellowish-green; shoots long, with long branches arising irregularly on the stem. Leaves 1.1–1.3 mm long, straight or somewhat curved to one side, widest near the base, gradually tapering to a long, narrow apex; nerve absent. Leaf cells, except at extreme base, long and narrow, with rather thin cell walls. At the leaf base is a small group of highly differentiated, large, rectangular, often strongly coloured, inflated alar cells. Fertile plants frequent. Seta long; capsule cylindrical, horizontal or inclined, operculum shortly rostrate.

Recognition The group of very large, inflated alar cells at the angles of the leaf base is an important feature of the genus *Sematophyllum*. The leaves of *Hypnum* also lack a nerve (or have a very short obscure one), but they are more strongly curved to one side or sickle-shaped and have small alar cells. *S. erythrocaulon* resembles *S. helenicum*, but can usually be recognised in the field by the more silky appearance afforded by its long slender branches and narrower leaves with long, narrow tips, and often by its frequent occurrence as 'hanging' mats. Microscopically, it differs in the leaves being widest near the base and often by the less thickened leaf cells. Plants intermediate in habit cannot be named in the field.

Habitats Found in open to deeply shaded places as 'hanging' mats on trees, decaying vegetation, steep rock faces and peaty soil banks. Also commonly growing on the trunks and branches of tree fern, black cabbage trees and other shrubs.

Status and distribution Apparently endemic. It seems to be confined to the Central Ridge and mainly to Diana's Peak National Park, where it is widespread and rather common. The genus *Sematophyllum* needs reassessment, at least in Africa, and a thorough revision may show that *S. erythrocaulon* is conspecific with known species from other tropical areas.

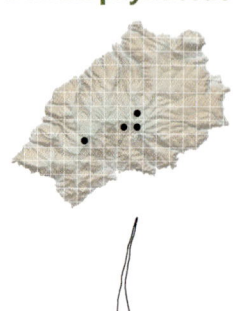

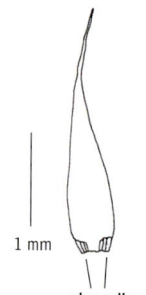

1 mm

alar cells

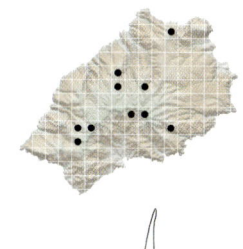

Moist shoots LEFT With sporophytes

Sematophyllum helenicum

Sematophyllaceae

Description Plants forming compact mats, the main stems creeping, with short, spreading or ascending branches arising irregularly along them. Leaves to 1.2–1.5 mm long, glossy, widest about $^1/_3$ the distance from the base, concave, straight or slightly curved, lacking a nerve; when moist, spreading widely from the stem, sometimes giving the shoot a rather spiky appearance. Leaf cells, except at base, long and narrow, with fairly thick cell walls; at the extreme leaf base (the alar region) is a highly differentiated group of large, rectangular, inflated cells (Fig. 8.19, p. 123).

Recognition The nerveless leaves with the group of large, inflated basal cells are characteristic of the genus. *S. helenicum* is usually differentiated in the field from *S. erythrocaulon* by its less silky and sometimes more spiky appearance, shorter branches and proportionately broader leaves. The leaf cell walls of *S. helenicum* are often thicker than those of *S. erythrocaulon*, though they are variable. Plants intermediate in character should be confirmed microscopically.

Habitats *S. helenicum* is a plant of shaded or open places, growing in mats on a wide range of non-native trees (including eucalyptus spp., acacia/wattle spp., Caffra thorn, white olive and Bermudan cedar), and on stumps, rotting wood, rocks and soil. In Diana's Peak National Park found only on an isolated Bermudan cedar tree, but perhaps overlooked on other hosts.

Status and distribution Apparently endemic to St Helena. It is widespread and locally rather common at middle to high altitudes, though seems to be rare in Diana's Peak National Park, where it is largely replaced by *S. erythrocaulon*. Like *S. erythrocaulon*, a taxonomic revision of the genus may show *S. helenicum* to be synonymous with species described from elsewhere.

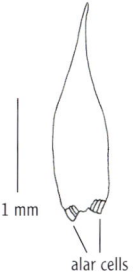

1 mm

alar cells

Abbreviations

adj. adjective
pl. plural
sp. species (single)
spp. species (plural)
subsp. subspecies

Glossary of technical terms

acrocarpous – in mosses: bearing the archegonia (therefore also seta and capsule) at the apex of a main stem or branch. Acrocarpous species usually have ascending or erect stems, form tufts or cushions, or occur as isolated shoots, and are sparingly branched (never pinnate) (Fig. 7) (compare with pleurocarpous).

acute – sharply pointed.

alar cells – cells at the basal outer angles of the leaf, differentiated in shape, size and/or colour from cells in the rest of the leaf, and sometimes forming ear-like projections (Figs 8.19 and 8.21).

antheridium (pl. **antheridia**) – male reproductive organ, usually globose to broadly cylindrical, that produces the antherozoids (sperm cells). (Fig. 4).

antherozoid – the motile male sex cell (Fig. 4).

apical lamina – in *Fissidens*, the part of the leaf extending beyond the sheathing lamina (in the half of the leaf that lies nearest the stem in situ, opposite the dorsal lamina) (Fig. 8.14).

appressed – lying close together, or closely pressed to the stem.

archegonium (pl. **archegonia**) – female reproductive organ (Fig. 4).

autoicous – with male and female inflorescences borne on the same plant, but on separate branches (Fig. 8.1).

axil (adj. **axillary**) – the angle between leaf (or bract) and stem.

bilobed – divided into two parts (e.g. Fig. 8.13).

bipinnate – a pinnate frond in which the primary branches are themselves pinnate.

bracts – leaves surrounding reproductive organs, sometimes differing from the other leaves in size, shape or form (Fig. 5).

bulbil – a bulb-like vegetative propagule, sometimes with tiny, leaf-like projections from the apex, developed in the axils of leaves (only in *Bryum dichotomum* in the St Helena bryophyte flora) (Figs 8.16 and 8.17).

calyptra (pl. **calyptrae**) – a membrane of tissue that protects the developing sporophyte: in mosses, it is carried upwards on the elongating seta and covers the developing capsule (usually conical or bell-shaped at maturity, sometimes with a 'beak', as in *Entosthodon*) (Fig. 7); in some thallose liverworts, it is a cylindrical, fleshy structure that encloses the developing capsule, and from which the seta and capsule emerge; in leafy liverworts, the calyptrae are usually inconspicuous.

capsule – the spore case, often differentiated into a spore-bearing urn and a sterile neck (Figs 6 and 7).

chloroplast – a body within the cell containing chlorophyll (green pigment) for photosynthesis (very small, except in hornworts: Fig. 6).

cladocarpous – in the pleurocarpous group of mosses, cladocarpous species bear the archegonia at the end of short branches (*Macrocoma* and *Macromitrium* in the St Helena flora).

comal – in acrocarpous mosses, referring to a group of usually larger leaves crowded at the shoot apex, and sometimes at intervals along a shoot.

cortex (adj. **cortical**) – referring to the outer or epidermal cell layer(s) of a stem (Fig. 5).

crenulate – with small, rounded projections, usually formed by bulging cell walls (often referring to leaf margins) (Fig. 8.10).

crisped – crinkled, rather 'wavy'.

cuticle – a non-cellular covering of organs including leaves and capsules.

deciduous – (of leaves) shedding, or falling away from the stem.

decurrent – leaf margins extending down the stem, forming a ridge or a narrowly triangular extension (Fig. 8.13).

dehisce – referring to a capsule splitting longitudinally or opening by an lid to release the spores.

dentate – strongly toothed: in leaves, with part of the cell projecting far beyond the leaf margin (or more than one cell projecting) (Fig. 8.12).

denticulate – finely toothed: in leaves, with only a small part of the cell projecting beyond the leaf margin (Fig. 8.11).

dioicous – with male and female inflorescences borne on different plants (Figs 8.2 and 8.3).

distant – (usually of leaves) well-spaced on the shoot, not crowded.

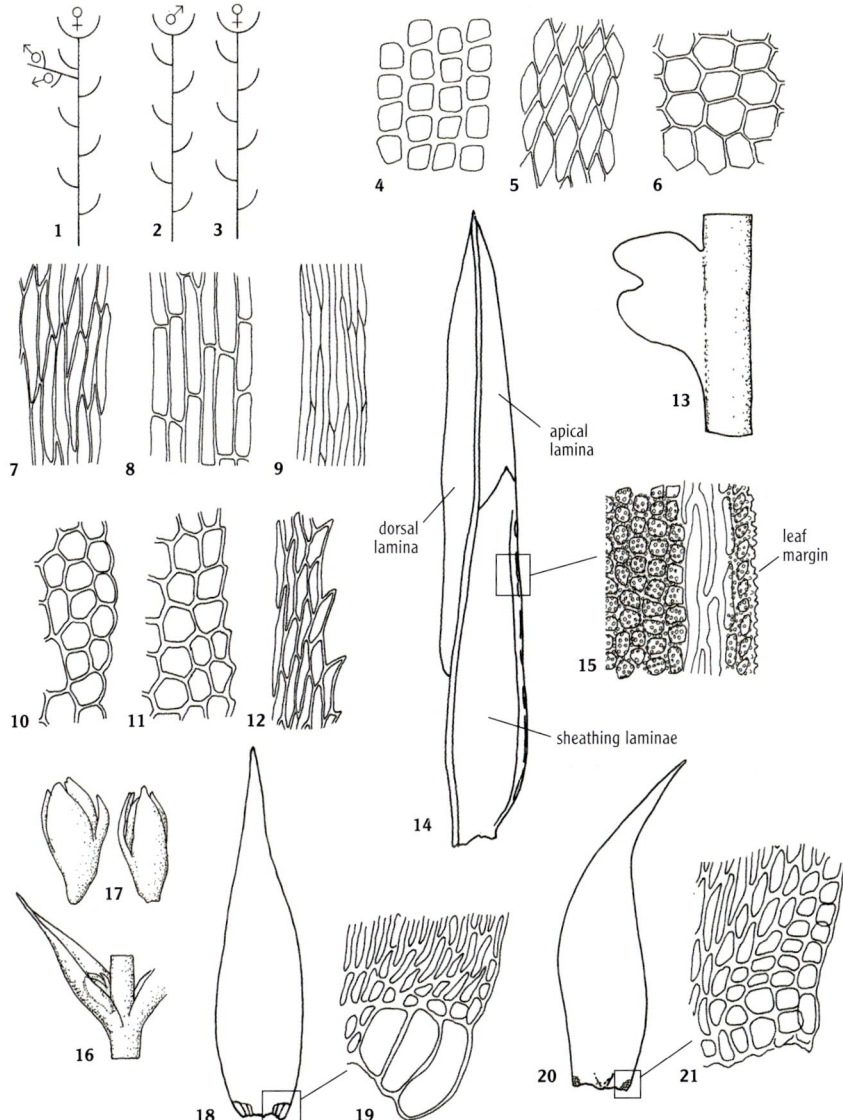

Figure 8. Spatial arrangements of male (♂) **and female** (♀) **structures on shoots**: 1, autoicous; 2, 3, dioicous.
Cell shape: 4, square; 5, rhomboidal; 6, hexagonal; 7, elongate hexagonal/pentagonal; 8, elongate rectangular;
9, linear. **Leaf margin**: 10, crenulate; 11, denticulate; 12, dentate, serrate. **Leaf of** *Cephalozia*: 13, bilobed leaf
showing decurrent base. **Fissidens**: 14, leaf, consisting of lower sheathing laminae, (i.e. two laminae, approximately
equal in size, folded closely against each other), an apical lamina above, and a dorsal lamina on the opposite side of the
nerve. 15, enlarged portion of intramarginal border (*F. chioneurus*). **Bulbils**: 16, bulbils in situ in leaf axil; 17, bulbils
enlarged. **Alar cells**: 18, leaf of *Sematophyllum helenicum*; 19, alar cells enlarged; 20, leaf of *Hypnum cupressiforme*;
21. alar cells enlarged.

dorsal lamina – In *Fissidens*, the part of the leaf on the 'outer' side of the nerve (in the half of the leaf that lies furthest from the stem in situ, opposite the sheathing and apical laminae) (Fig. 8.14).

epiphyte (adj. **epiphytic**) – a plant growing on another plant (here usually referring to bryophytes growing on bark).

excurrent – referring to a nerve projecting beyond the leaf apex.

flagelliform – referring to a long, slender, tapering stem or branch with leaves progressively reducing in size towards the tip.

flexuose – moderately wavy or twisted.

gemma (pl. **gemmae**; adj. **gemmiferous**) – a small globose to cylindrical or discoid, few-celled body serving in vegetative reproduction Fig. 5).

glaucous – with a greyish, bluish or whitish bloom.

globose – spherical in shape, or nearly so.

hexagonal – of cells, six-sided (Fig. 8.6).

hyaline – colourless and transparent.

immersed – covered or hidden; immersed capsules are overtopped by the perichaetial leaves and are partly hidden.

inclined – of capsules, held on the seta at an angle between erect and horizontal.

incurved – curved upwards and inwards (usually referring to leaf margins).

inflorescence – the male or female reproductive organs (antheridia or archegonia) together with the surrounding bracts.

inrolled – rolled inwards, denoting a greater degree of curling inwards than 'incurved'.

intramarginal – referring to a leaf border (only in *Fissidens* in this guide) in which a zone of elongate cells is separated from the leaf margin (outer edge of sheathing lamina) by 1 or few rows of square or rounded laminal cells (Fig. 8.15).

involucre – a tube-, scale- or cup-like structure protecting the male or female reproductive organs, or the developing sporophyte.

keel – the folded lower edge of the lobule in a leafy liverwort (Fig. 5); a longitudinal ridge on the perianth of a leafy liverwort; the ridge on the back of a partly folded moss leaf (like the keel of a boat).

lamina (adj. **laminal**) – the blade part of the leaf (excluding the nerve).

lanceolate – of leaves: shaped like the head of a lance, narrower than 'ovate' (Fig. 7).

linear – of cells, long and narrow (e.g. Fig. 8.9).

lobule – a smaller segment of a leaf that is at least partially folded against a larger segment. Lobules are found in many leafy liverworts where the leaves are divided into two unequal portions (Fig. 5).

Macaronesia – collective name for the groups of islands in the north Atlantic Ocean which include the Azores, Canary, Cape Verde, Madeira and Savage Islands.

mammilla (pl. **mammillae**; adj. **mammillose**) – a hollow, bulging projection from the cell surface into which the central cell cavity extends (cannot be distinguished from a papilla in surface view).

medulla (adj. **medullary**) – the inner tissue (cells) of a stem (Fig. 5).

microphyllous – bearing leaves that are very small compared to normal leaves (usually they are found on stems or branches that are much smaller than normal).

monoicous – with male and female inflorescences borne on the same plant (compare with autoicous, dioicous).

oblique – at an angle.

oblong – rectangular, with rounded corners.

oil body – an oil-rich body in the cells of most liverworts, usually colourless, often providing a valuable aid to identification.

operculum – lid of a capsule (Fig. 7).

obtuse – blunt.

ovate – (of leaves), widest below the middle of the leaf, broader than 'lanceolate' (Fig. 7).

ovoid – an egg-shaped 3-dimensional body.

papilla (pl. **papillae**; adj. **papillose**) – a small, solid projection from the surface of a cell (cannot be distinguished from a mammilla in surface view).

pendent (= **pendulous**) – hanging.

perianth – a thin, broad or narrow, tube-like structure enclosing the archegonia in most leafy liverworts, from which the capsule and seta emerge at maturity.

perichaetial leaves – the leaves (or bracts) immediately enclosing the female parts (the archegonia), therefore surrounding the base of the seta. Perichaetial leaves are often longer than the leaves below them on the stem.

peristome – the ring of teeth that surround the mouth of the capsule in most mosses.

petiole – the leaf-stalk of a vascular plant or fern.

pinnate – usually referring to shoots in which the branches arise in an almost regular pattern on opposite sides of a stem, lying in one plane, so that the frond often looks feather-like (see also bipinnate).

pleurocarpous – in mosses: bearing the archegonia (therefore also seta and capsule) in small lateral buds on stems or branches. Pleurocarpous species are typically mat-forming, have creeping, prostrate main stems and are freely branched, often pinnately so (Fig. 7) (compare with acrocarpous).

procumbent – prostrate; plant lying nearly flat on the substrate.

prothallus – (of ferns), a small (3–20 mm), green, rounded disc-like or sometimes strap-like body, one cell thick, in which the male and female reproductive organs develop.

protonema – a minute, filamentous (rarely thallose) structure, developed from the spore following germination, and from which the green plant (gametophyte) develops (Fig. 4).

pyriform – pear-shaped.

receptacle – in thallose liverworts, a disc or wart-like structure bearing the archegonia, borne on the thallus or raised on a stalk (Fig. 6).

rectangular – of cells, longer than wide, with straight sides (e.g. Fig. 8.8).

recurved – curved downwards or backwards.

reflexed – bent abruptly backwards.

rhizoid – a thread-like outgrowth that functions as an anchor (but some bear propagules (tubers) that function in vegetative reproduction). Rhizoids are unicellular and usually colourless in liverworts, multicelluar and often coloured in mosses.

rhizome – a creeping, horizontal, often underground stem.

rhomboidal –diamond-shaped (can be short or elongate) (Fig. 8.5).

rostrum (adj. **rostrate**) – a beak (e.g. on a calyptra, or on the operculum (lid) of a capsule, or at the apex of a perianth).

serrate – saw-toothed, with teeth projected forwards (Fig. 8.12).

seta (pl. **setae**) – stalk bearing the capsule.

sheathing lamina – In *Fissidens*, the 'double' (folded) lamina forming the lower 'inner' part of the leaf, its base(s) clasping the stem, and above, partially sheathing the adjacent leaf (Fig. 8.14).

spreading – referring to leaves diverging from the stem at a wide angle (see also spreading-erect)

spreading-erect – referring to leaves diverging from the stem at an appreciable but moderately narrow angle.

spinose – spine-like, with one or more spines.

spore – in bryophytes, a single-celled, non-sexual propagule formed in the sporophyte capsule that is dispersed usually by the wind and develops into the gametophyte (i.e. the green leafy or thallose plant we recognise as a liverwort, hornwort or moss).

square – of cells, approximately equal in length and width (Fig. 8.4).

stylus – a strap- or club-shaped structure found between the lobule and the stem in certain leafy liverworts (see *Cololejeunea grossestyla*).

sub - a prefix denoting nearly, almost (e.g. in sub-erect, sub-rectangular).

thallus (adj. **thallose**, **thalloid**) – a flattened plant body, not differentiated into stem and leaves.

tomentum (adj. **tomentose**) – a woolly mass of branched filaments.

transverse – across; perpendicular (at 90°) to the axis; often used to indicate a cross-ways insertion of the leaf on to the stem (e.g. in *Cephaloziella*).

trigone – a 3-angled thickening at the corner of cells (Fig. 5).

truncate – abruptly cut off or ending abruptly.

tuber (adj. **tuberous**) – a rounded or ellipsoid body borne on rhizoids (usually subterranean), and serving in vegetative reproduction.

underleaves – in some leafy liverworts, a third row of leaves along the underside of the shoot, usually smaller and differing in shape from the other leaves (the lateral leaves).

undulate – referring to a margin or surface that is wavy up and down.

zygote – a very young embryo resulting from the union of male and female sex cells, the zygote eventually developing into the sporophyte (Fig. 4).

Further reading

Hooker, J. D. & Taylor, T. 1845. Hepaticae Antarcticae supplementum, on specific characters with brief descriptions of some additional species of the Hepaticae of the Antarctic regions, New Zealand and Tasmania, together with a few from the Atlantic islands and New Holland. *London Journal of Botany* 4: 79–97.

Mitten, W. 1875. Musci (pp. 357–366), Hepaticae (pp. 366–373), In: J.C.Melliss, *St. Helena. A physical, historical and topographical description of the island, including its geology, fauna, flora and meteorology.* London: L. Reeve & Co.

Müller, F. 1999. Contribution to the bryoflora and bryogeography of St Helena (South Atlantic Ocean). *Tropical Bryology* 16: 131–138.

Online references

An online book that includes a great deal of information on the structures and life cycles of bryophytes, and covering a wide range of ecological topics:
Glime, J. http://www.bryoecol.mtu.edu/

Useful general information on structure and life cycles, and on ecology and classification:
Australian National Botanic Gardens. http://www.anbg.gov.au/bryophyte/

Appendix:
Flowering plants, conifers and ferns mentioned in the species accounts

acacia/wattle Includes various species: *Acacia melanoxylon* (blackwood) is common in the uplands, with *A. mearnsii* (black wattle) and *A. longifolia* (willow) at gradually lower altitudes
Bermudan cedar *Juniperus bermudiana*
bilberry tree *Solanum mauritianum*
black cabbage tree *Melanodendron integrifolium*
buck's-horn *Lycopodiella cernua*
Caffra thorn *Erythrina caffra*
Cape yew *Afrocarpus falcatus*
creeper *Carpobrotus edulis*
Diana's Peak grass *Carex dianae*
eucalyptus At least 10 species found on the island: especially *E. grandis* (great gum)
filmy fern *Hymenophyllum capillaceum*
he cabbage *Pladaroxylon leucadendron*
Mexican cypress *Cupressus lusitanica*
Monterey cypress *Cupressus macrocarpa*
New Zealand flax *Phormium tenax*
Norfolk Island pine *Araucaria heterophylla*
rebony *Trochetiopsis* × *benjaminii*
rose apple *Syzygium jambos*
small fuchsia *Fuchsia coccinea*
small kidney fern *Dryopteris napoleonis*
soft rush *Juncus effusus*
tongue-ferns *Elaphoglossum conforme* (common), *E. nervosum* (veined) and *Pleopeltis macrocarpa* (dotted tongue-ferns) grow epiphytically on the Peaks
tree fern *Dicksonia arborescens*
white olive *Elaeodendron croceum*
whitewood *Petrobium arboreum*

Index

Names in **bold** type are those currently accepted as valid for the species. All other names are synonyms.